VALLEY OF EDEN

edited by

Anthony Austin

9 780954 029845

VALLEY OF EDEN

edited by Anthony B Austin

Published by-

Black Rabbit Press
15 Homewood
Findon
West Sussex
UK BN14 0XA

blackrabbitpress@hotmail.com

+44 (0)1903 877 937

Published 2008. First Edition
Printed by Lightning Source

ISBN 978-0-9540298-4-5

9 780954 029845

Some pre-publication comments-

"Connects Armenia, New Zealand, and the planet Mars in a totally unexpected manner... takes your breath away with its almost offhand revelation of masonic style secrets"

"Another we-dare-you-to-believe-it book from Black Rabbit Press"

"Brilliant rewrite of the civilising of humanity by the teachers from space starting in Armenia and finishing with a part-Aryan tribe one whisker from reversion to a pre-Adamic stone age society"

"Tightly written narrative that removes many commonly held illusions concerning the last few thousand years...too short by a megalithic mile"

"Khatchvar pielidismacraeli Karrevanrath, dlo?"

ACKNOWLEDGEMENTS AND THANKS

'Valley of Eden' was assembled over a period of six years beginning in 2001, from data supplied by a number of individuals who the Editor wishes to thank, for without their help and encouragement, this work would not exist. Most contributors are identified in the text where possible or in the Appendix, but there are some who prefer to remain nameless or unidentifiable for reasons of their own. The Editor nevertheless apologises to all persons whose religious sensibilities are offended by any part of this work, which is intended to be non-denominational and factual, taking complete responsibility for the work of editing. The Editor also apologises for the less than perfect quality of this book, which was hand produced on a vintage word processor running on sweat and tears.
A special mention and davenning for Brian Crowley who is ill in hospital at the time this book goes to press. Best wishes and sympathies from everyone.

SOME PREVIOUS WORKS

'The New Life Prophecies of Frances Yule' - Ed. Venestus, Destiny Publications NZ, South Australia, 1982.

'Return to Mars' - Crowley and Pollock, Magistra, Victoria, Australia, 1989 and Black Rabbit Press, West Sussex, UK 2003.

'The Dragon's' Tail' - Austin and Crowley, IllumiNet, Ga. USA, 2000.

'The Sand Whales of Mars' - McLeod and Austin, Black Rabbit Press, West Sussex, UK, 2001.

'Mother Shipton: The Missing Prophecies' - Austin and Yule, Black Rabbit Press, West Sussex, UK, 2003.

'Draco: The Tenth Planet' - Austin and Crowley, Black Rabbit Press, West Sussex, UK, 2003.

Articles on Prophecy: Austin, www.prophecies.us 2006.

'The Branson Letter' - Austin, www.enigmasmania 2007.

VALLEY OF EDEN

PREFACE

Mars, Armenia, and New Zealand. What possible connection could there be between these places? Does this book belong on the science fiction shelf instead of the historical section? Where, for that matter, is Armenia and should we care?

The first point we wish to make is that Armenia is certainly a small insignificant country today, but what if it was chosen by the gods of Mars because of the superior intelligence of its people, who could be taught? What if there was evidence of one-time Martians on Earth in Armenia, of all places?

And consider the Maori, the native New Zealander, who has detailed stories of how his ancestors emigrated from Armenia, taking with them a knowledge of exactly who they were, Hawaiki notwithstanding, and claiming to be part Aryan?

What if there were monuments in Armenia that have exactly the same shape as some standing stones of Mars... would you wonder why? Would you think it possible that extraterrestrials much like ourselves could instruct our ancestors about how Mars used to look, ages ago, when it still had an ocean, making sure this information would be remembered thousands of years later? Quite a clever trick, wouldn't you think?

Perhaps you consider these items to be part of a spoof, or a joke, aimed at the gullible and the credulous, but that is not so. Our evidence is gathered from several sources now considered credible, such as early Hebrew texts, and the Zoroastrian religion whose origin lies so far in the past that nobody knows when it began. Stories from the past, yes. Myths, no.

It is not that we mind criticism; the authors have attracted so much of that over the years that we have become impervious to mud slinging, which always misses its target, often because the critic cannot be bothered to read the book! We admit that much work remains to be done but that must be taken on by others because the 1001 authors and the editor are 'elderly' so while the spirit is willing... you know the rest.

We do not intend to teach or to give lectures, and do not care how much money this book makes, if any, so long as minds are opened and the right questions are asked in the right places, which means politics and religion.

All that will be said at this place is that the ancient Armenians were absolutely obsessed with Mars, tiny planet that it is, because they littered the headwaters of mountain streams with martian symbols including stone rams that followed paleolithic cave art including perfectly formed signs of Aries painted on cave walls. Although Armenia was one of the first countries to accept Christianity, certain stone objects that were necessary to earlier religions were retained, even in churches, in case Christianity turned out to be no more than a passing fad. Church walls were decorated with pre-Christian heroes such as King David with his sling taking on Goliath, with a sleeping ram at his feet that would in all likelihood have been painted red, the colour of the martian sand. There is a red ram in Genesis!

Above the ram are two symbols, both flowers, that symbolise the Garden of Eden which is now admitted by Hebraic authorities to have been in Armenia, in or near Yerevan, the capital city. We found that out for ourselves in 2001.

NON HUMAN SKULL FROM PERU

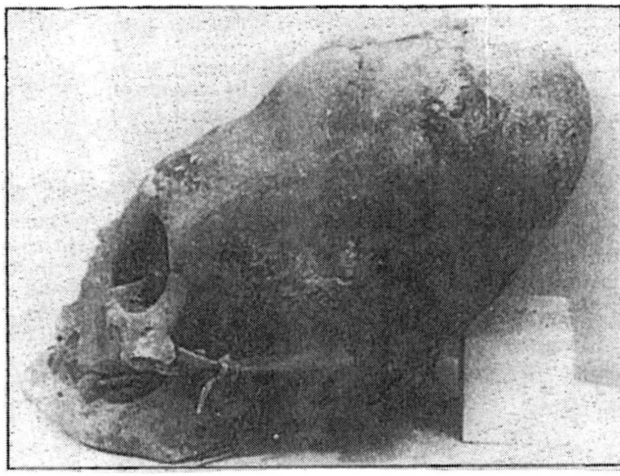

> *"It is claimed that these skulls were those of "rulers." In any event, the notion that the deformity is natural or genetic is buttressed by the discovery of similarly shaped skulls in the fetuses of mummies."*

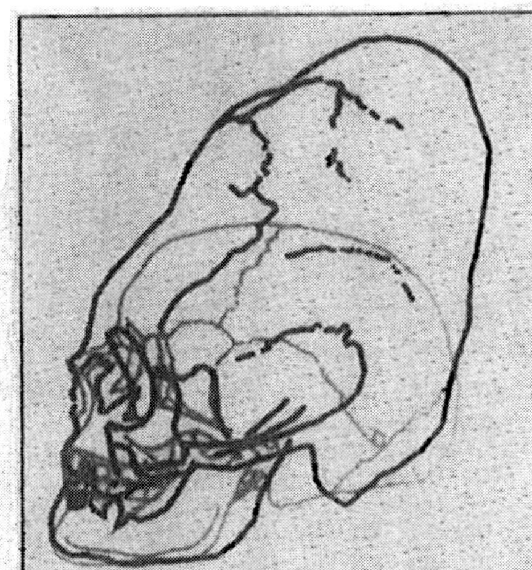

Top: Elongated skull from museum in Lima, Peru. Bottom: Drawing by Lumir Janku showing contrast with a normal skull. Photos by Bob Connolly.

VALLEY OF EDEN

Ahura Mazda spoke unto Spitama Zarathustra saying: I have made every land dear to its peoples even though it had no charms whatsoever in it; had I not made every land dear to its peoples even though it had no charms whatsoever in it, then the whole living world would have invaded the Airyana Vaejah.

Fargard 1. Trans. James Darmesteter
from Sacred Books of the East,
American Edition, 1898.

Form divisions in the Garden of Eden

The Book of the Vestment

VALLEY OF EDEN

This work will attempt to answer some age-old questions concerning a place known to mankind as the Garden of Eden and Paradise; the two terms being thought interchangeable, although strictly speaking, the Garden of Eden was located in the east of Paradise according to the Bible, roughly where the city of Yerevan lies in the country of present day Armenia.

Commentators on the Book of Genesis, where this tale is to be found, have often mistakenly located the Garden of Eden, or Eden for short, in several other locations on Earth, ignoring all evidence to the contrary including the fact that Paradise was known as Dilmun and early geographers placed Dilmun in Armenia. We shall provide that evidence later, meanwhile noting that Dilmun was and is the mountainous area overhanging the plains of Gilan, or present day Azerbaijan.

Armenia is located to the east of Turkey, to the west of Azerbaijan, south of Georgia, and to the north of Iran/Iraq at about the 40th parallel and is generally at about 2000 feet altitude, enjoying a temperate climate. The highlands are volcanic, hence the soil is fertile, and it is probable that Armenia has never been completely glaciated even during the last Ice Age, for stone tool production continued even on the slopes of 15,000 foot high volcanos for hundreds of thousands of years.

Map of Armenia showing Eden in the middle of Paradise.

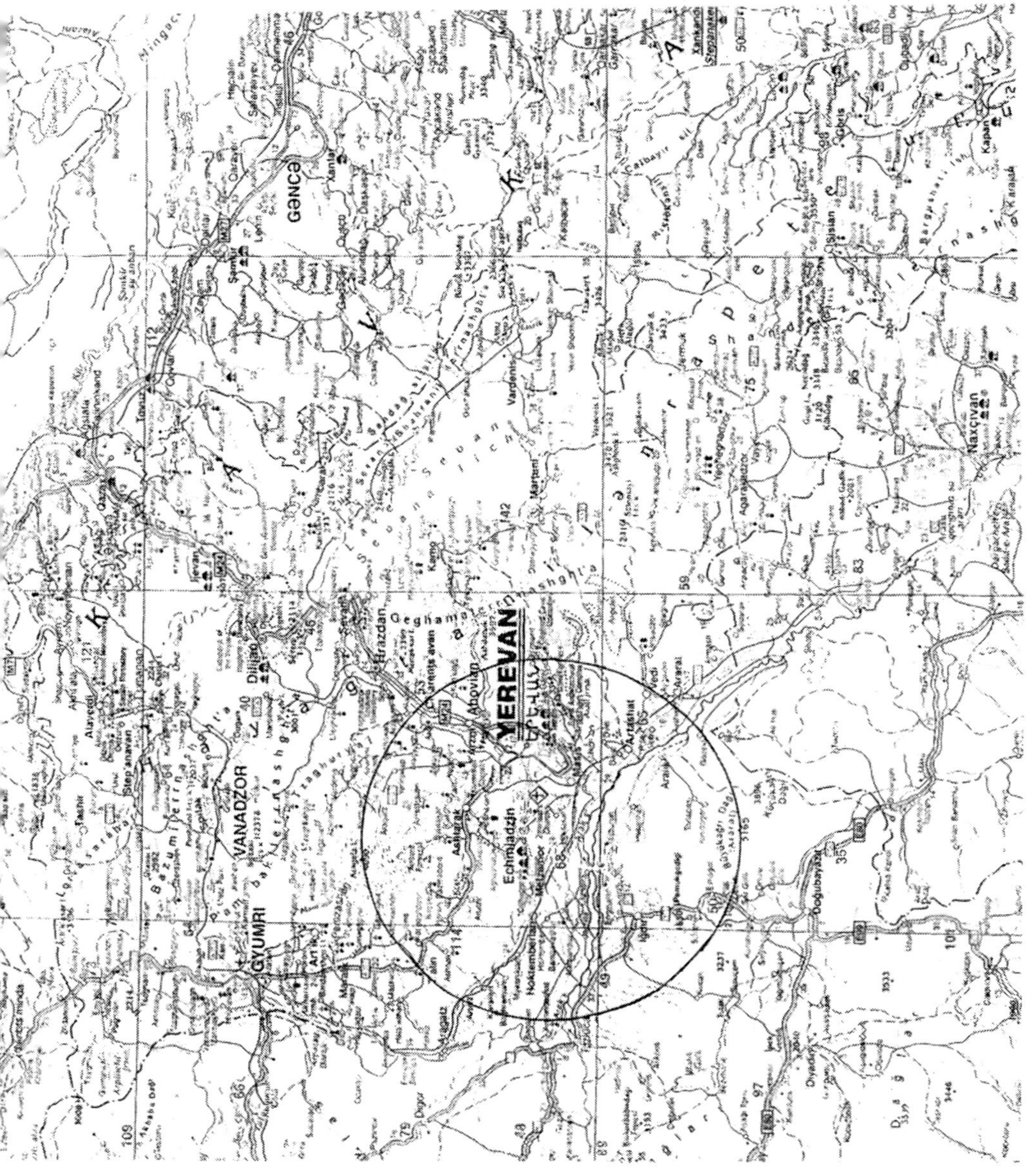

Part of Armenia and Turkey
with Paradise shown as a circle

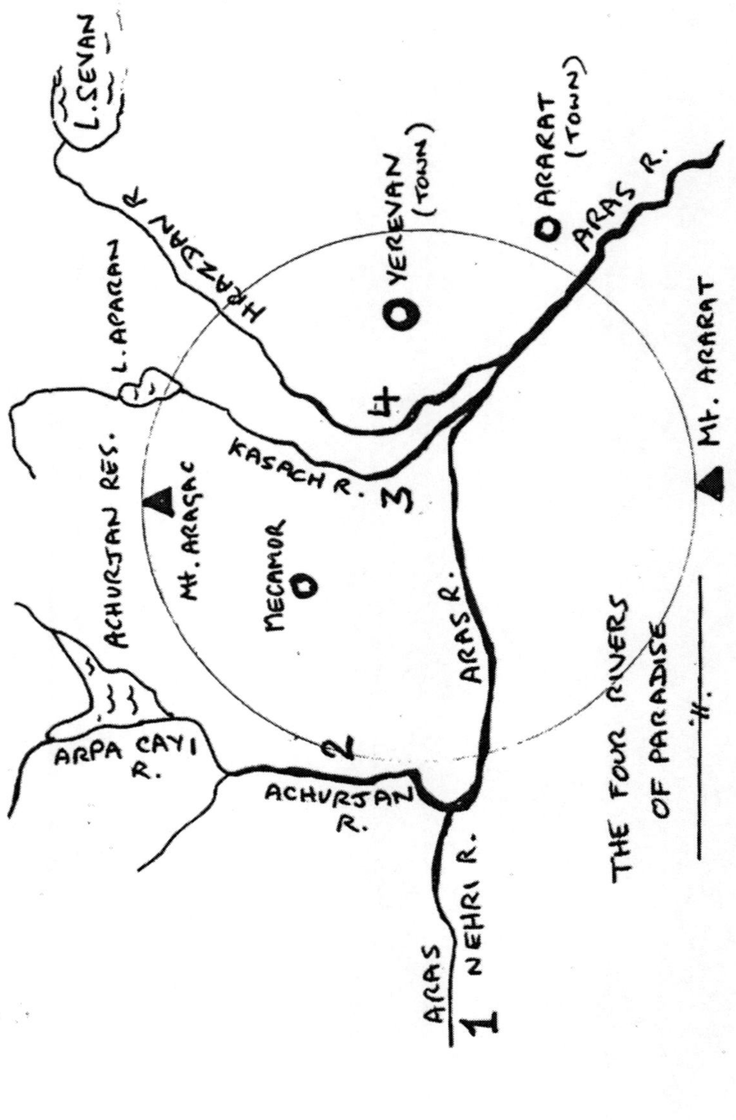
L. SEVAN
HRAZDAN R.
L. APARAN
YEREVAN (TOWN)
ARARAT (TOWN)
ARAS R.
Mt. ARARAT
KASACH R.
4
3
ACHURJAN RES.
Mt. ARAGAC
MECAMOR
ARAS R.
ARPA CAYI R.
2
ACHURJAN R.
ARAS NEHRI R.
1
THE FOUR RIVERS OF PARADISE

The major river of Paradise is the Aras (or Araxes) flowing through a gorge which widens out near Yerevan to form a plain suitable for agriculture, and then flows south, joining the Kür river before emptying into the Caspian Sea. The Aras is not named in Genesis but it was said that there were four rivers of Paradise, giving commentators on Genesis a real headache as to their identity and location. Here, we prefer to identify the four rivers as the Hrazdan, Kasach, Achurjan, and Aras Nehri rivers, ignoring those rivers mentioned in Genesis (Euphrates, Pison, Tigris, Gihon) which merely define the location of the land of Armenia, and none of them flow through or near the Garden of Eden. This will upset a *lot* of people but we shall present our evidence as we proceed.

Armenia and Georgia have the Caucuses Ranges, which are admitted to having been the cradle of the white or Caucasian races and the home of the Indo-European languages. It is a popular myth that these white folks mutated from incoming African blacks, which is demolished at one stroke by noting that the gene for black skin is recessive in that repeated mating of black with white eliminates the black gene. From where, then, did the gene for the Caucasian race come from?
It might as well have come from space for all we know, but the solution involves the 'impossible' explanation that the human race is much older than is realised, conquered space travel and other worlds so many years ago as to make discussion of that equally impossible, and visits from time to time or perhaps *all the time*, if 'they' live in the Solar System and *always have done*.

The origin of the Caucasian strain lies beyond the last Ice Age, but how many years beyond is impossible to answer, except scientists guess or estimate over 400,000 years. That is long enough for humans to have gone into space several times, either through their own efforts or with the assistance of other humans from other worlds. Examples abound of humans who were taken to 'heaven' but once gone, they seldom return to tell their tale, and if they do, it is not believed although it may be recorded

Without a doubt, metallurgy or the art of working metals began in the Caucases, centered in Armenia on the town of Mecamor (Mezdamor) where over 200 furnaces have been discovered, enabling the workers to produce some 15 different kinds of bronze, gold ornaments, and even manganese steel, starting many thousands of years ago. The tools and weapons that forged here enabled the Armenians to trade with their neighbours as far as a thousand miles away. These and other achievements are dealt with later in more detail in this work where the opinions of a contemporary Armenian are recorded.

The origin of the advances in technology of the Armenians 6000 years ago needs clarification and explanation; sadly lacking in conventional histories, and it is here that we must turn to religious texts along with surviving mythology and legends which explain that humanity had help, and it came from somewhere outside Earth. It came at the earliest possible opportunity when the climate had ameliorated from the frigid conditions of the last Ice Age and there were enough humans around to make this experiment, for that is what it was, worthwhile in the eyes of the teachers. As it happened, some inhabitants of Paradise, characterised by Adam, were not ready to absorb what was taught.

The origin of the teachers is stated in this work to be Mars, or that was their last port of call, which is in fact the only half-habitable world in the Solar System apart from Earth, which it once resembled. Certain details of Mars as it was have come down to us which were not known and could not have been known before the second half of the 20th century.

The inhabitants of Paradise were taught facts including the information that there had been only one ocean on Mars; that it was salt and extremely rough; that vapours arose and obscured parts of the landscape including the one central island of that ocean, which was named as was the sea, known to have been in the north. The colonists from Mars are identified. We know the name of the ship, the builder, and its captain, and the number of crew members. We know its shape and what they used for cosmic ray protection. Above all, we know where it landed. The crew are characterised as gods but appeared human, although slightly different to contemporary Cro-magnon, having fair curly hair and blue eyes, which had never been seen on the Planet until they arrived.

Should it be said that the Moon is closer and more likely as a staging post, this is possible but the legends say otherwise. The Moon is simply a mine which is still in the process of being dug up for metals and He3, used for power production. Much of this mining goes on in craters because these offer a good exposure of the bedrock and any minerals it may contain. Helium 3 is extracted by surface skimming and heating. The only thing Moonies need from Earth is water which the Moon lacks, although it has ice.

The so-called gods taught many different arts and sciences, such as astronomy, astrology, crop husbandry, their ideas on religion, stone-work, law, and other subjects that have been adequately covered in other books, particularly those by Sitchin. While we were worshipping the Sun and the Moon, they worshipped the stars such as Orion, Sirius, and the Plieades, passing on their useless lore to us including the crazy idea that when we died we might go to the stars as they did.

This symbol of the six pointed star is commonly thought to be the so-called Star of David although the Bible makes no reference to it. The earliest Jewish design incorporating the Star dates from the 11th century AD. This, or something like it, was used as an identifying mark by the 3rd Reich when it was usually yellow. Today it is used on the Israeli flag when it is colored light blue, dating from around 1948.

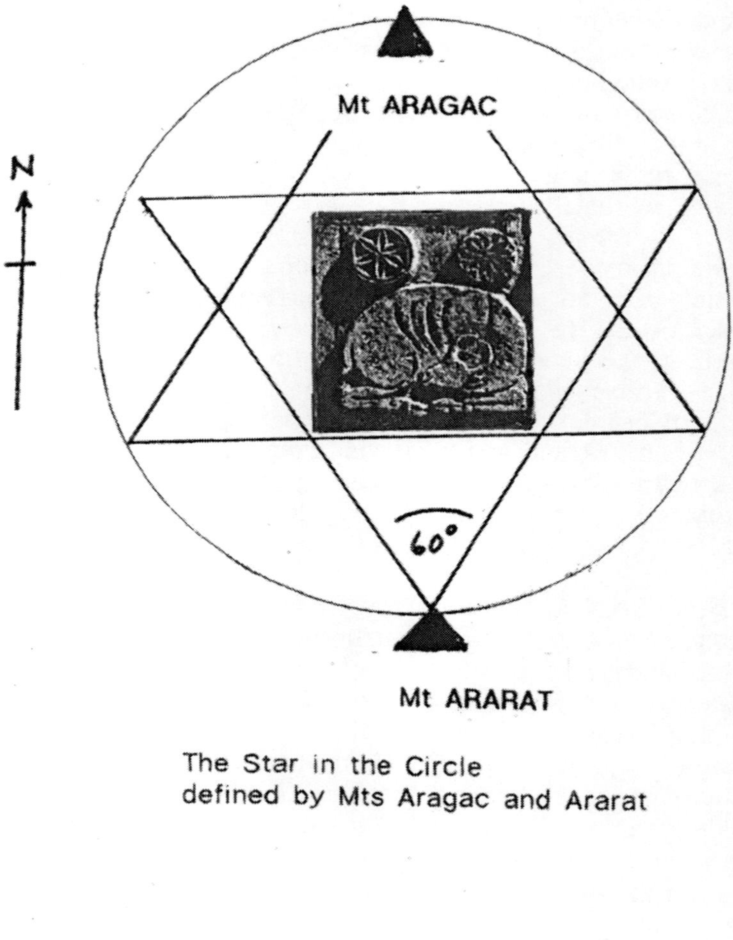

The Star in the Circle
defined by Mts Aragac and Ararat

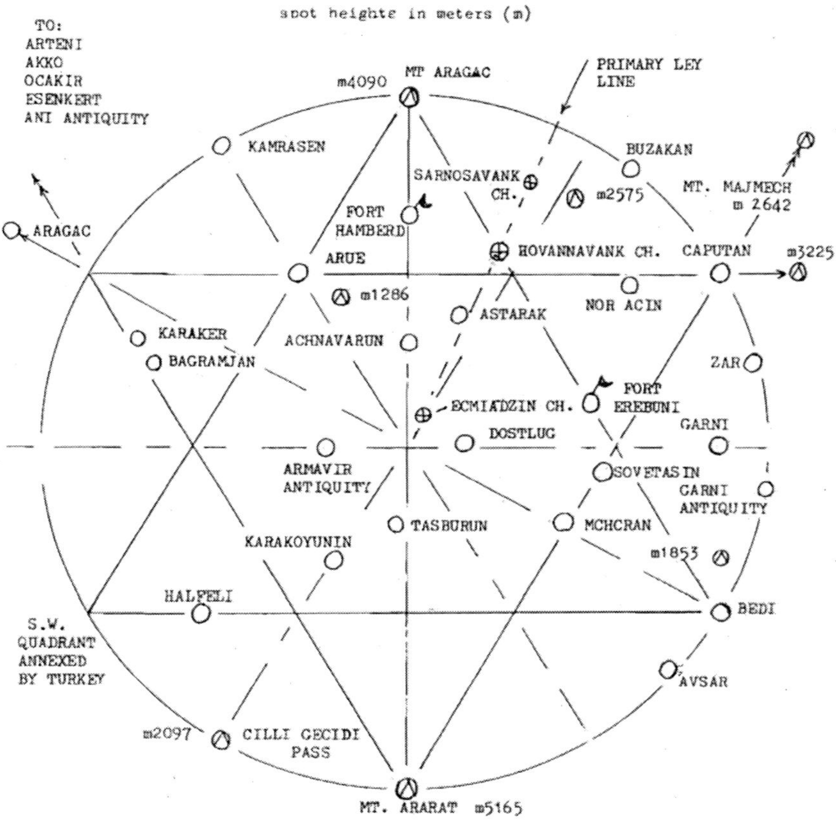

THE GATES OF EDEN

Some of the 50 plus correspondences with actual locations.
The primary ley-line sets out the 23.5 tilt of the Earth
c. Austin 2003

See also Appendix for more 'coincidences'

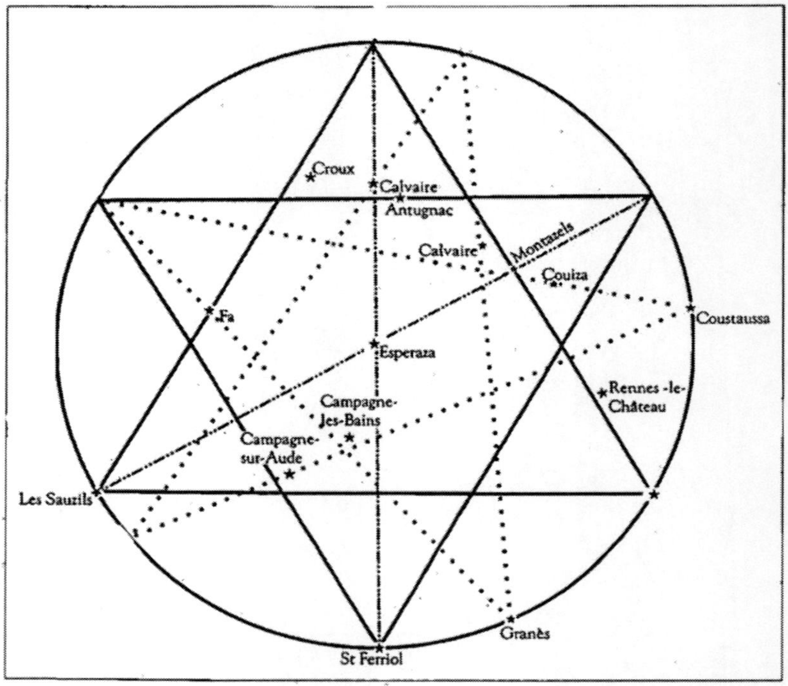

The five and six pointed stars superimposed on the landscape near Rennes-le-Chateau, France.

From works by Baignent Leigh & Lincoln

To ensure that the geography of Mars was remembered, clues were left all over the landscape of Paradise in the form of stone circles, other constructions of megalithic size, and standing stones; some with holes pierced through them like gun-sights. The object of this stone-heaving was to construct a giant circle in Paradise using the obvious markers for north and south of the two largest volcanos in the area, which are the Gates of Eden. A six pointed star was laid out within the circle along with at least three 'ley-lines' of doubtful purpose except for one which lays out the tilt of the Earth at 23.5 degrees. Another might be very close to being accurately East-West. This shows the circle was constructed fairly recently so that the north pole was then where it is now, more or less, and the author does not have facilities for measuring such things to a fine degree of exactitude, leaving that for another. Spherical geometry is too hard! The six points of the Star, as we shall call it, are echoed everywhere on Earth, including on the Israeli flag.

The colonists (Uruk in Phoenician) certainly left their genes behind on Earth but some may have returned from whence they came, leaving us with an unforgettable legacy of superior technology which we immediately exploited, not realising that the pristine beauty of the Garden of Eden would thereby be marred by mining for gold and metals which would be used in the forging of better weaponry. This resulted in the emergence of a rag-bag collection of evils that would be called Original Sin, for which we are still beating our breasts and seeking forgiveness from some Avatar, Messiah, or deity to this day. What we want is a return to the halcyon days of the Golden Age, long lost in the beautiful legends of the past, and for our consciences to be wiped clean, somehow, soon.

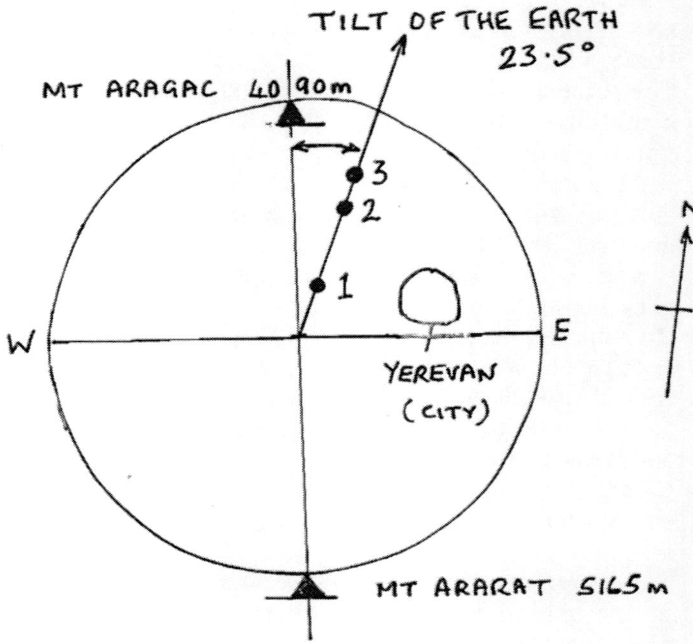

1 ECMIADZIN CHURCH
2 HOVANNAVANK CHURCH
3 SAMOSAVANK CHURCH.

ARMENIAN LEY LINE DISPLAYING 23.5°

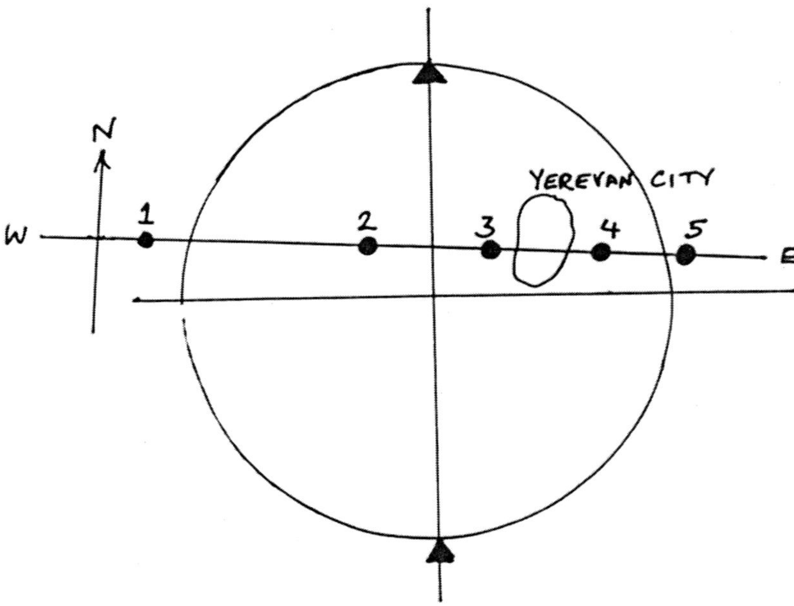

THE EAST-WEST LEY LINE OF PARADISE

1 BAGARAN
2 MECAMOR PREHISTORIC SITE
3 SVARTNOZ CHURCH
4 EREBUNI FORTRESS
5 GEGHARD CHURCH

And so it was that more emphasis was given to martial deities such as the god who was to later become known as Mars, who began as an agricultural deity only to end as the spawn of fear, terror, panic and evil who could not care less what grew in the Garden of Eden.

Although he was first worshipped as Ar, the Sun, he morphed into the deity Ara of the Armenians, becoming Ares of the Greeks and then Mars of the Roman Empire. He was imported into Scandinavian countries where he became Tiw, who is remembered in the second day of the week, Teusday, and his number is two, as in deuce. We are still pagans, it seems, counting Moon-day, Mars-day, Woden-day, Thor's-day, Freya-day, Saturn-day, and Sun-day.

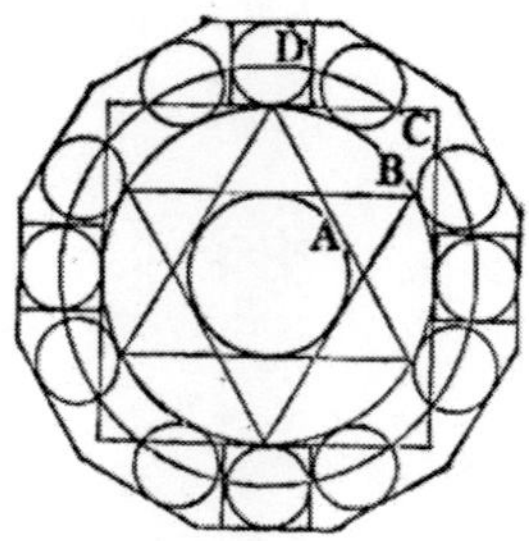

The Holy City according to J. Michell
from: City of Revelation, Abacus, p.81
The letters refer to its dimensions.

THE CALENDAR

Through the unknown remembered gate
When the last of Earth left to discover
Is that which was the beginning

Little Gidding - T.S. Eliot

THE CALENDAR

In this chapter, we shall deal with the objection that the Star in the Garden of Eden is not a reality in that one may prove anything if one draws enough lines! This is compounded by the fact that early maps of Armenia do not show all the places indicated in the drawings provided in this book, nor are those places in the same locations! Often, villages are shown on those earlier maps that seem to have disappeared or possibly changed their names, making correlation extremely difficult, compounded by the fact that early geographers were working without any survey tools and may not have even visited the places they were so confidently marking on their maps. The earlier we go back in time, the worse the maps until they could be called fictitious or imaginary. It is unnecessary to give examples for all these maps are good for is raising a smile of superiority on our faces.

Nevertheless, it is germane to point out that the markers of Mt Aragac and the two mountains of Greater and Lesser Ararat were known in the furthest antiquity and are shown on all maps of Armenia. If we should seek for the most ancient markers of Armenia, these would consist of upright stones installed at a time beyond any maps, some with 5 to 7 cm holes drilled through them, which were obviously sighting stones, and there are so many of them, amounting to hundreds or even thousands throughout Armenia that it is difficult to imagine that they had no purpose. The objection that diagrams in this book contain too many lines is therefore over-ruled, particularly when it is realised that the early Armenians certainly were astronomers and all ancient cultures were interested in circles that took ages to construct, eg Stonehenge.

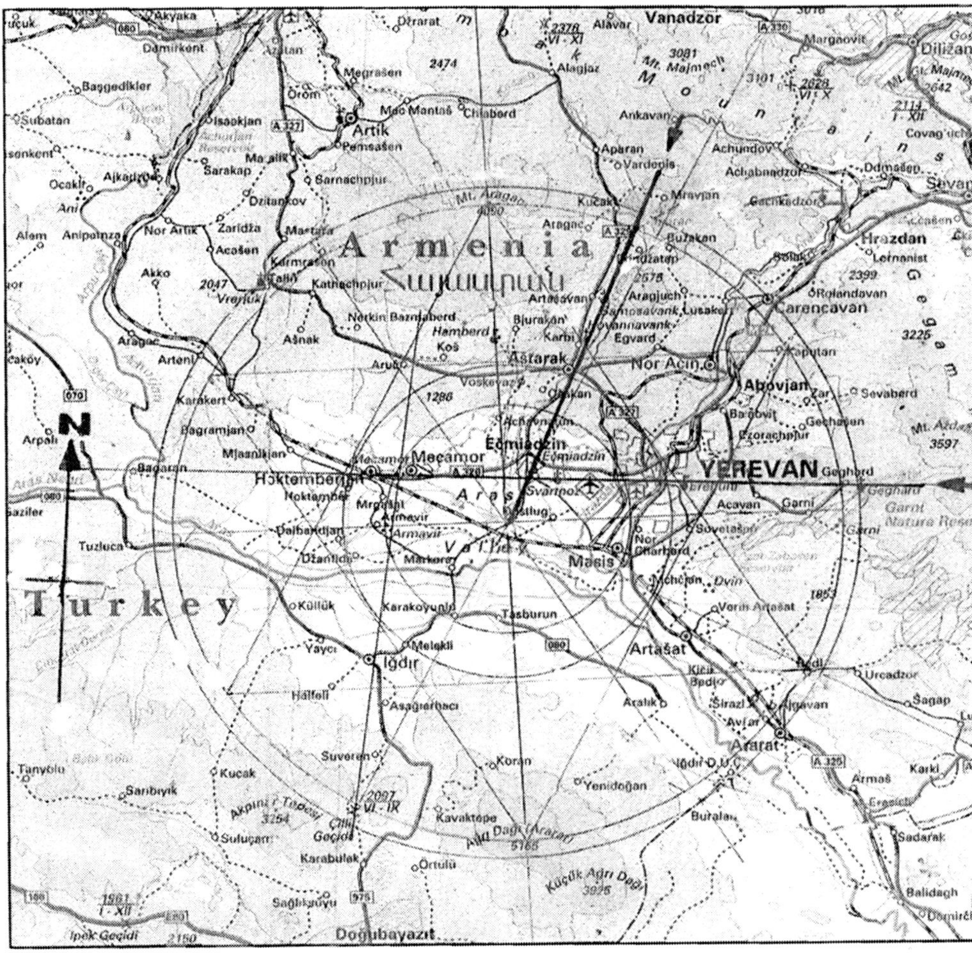

Several concentric circles drawn on a map of Armenia centered on the 'Aras Valley' with Mt Aragac to the north and Mt Ararat to the south. The heavy lines show 'ley lines'. For more detail and sites see other diagrams

would pass through the ancient megalithic sites of Armavir and Mecamor, but little else. Another circle could be drawn outside the circle containing the Star that would pass through (going clockwise from Aragac) Kucak, the major town of Carencavan, Geghard Church, the ancient town of Ararat, Ateni, and the church of Talin. Taken together, these three circles are repeated in any number of diagrams from the most ancient times to the present day as though it was some kind of race memory and 'the way things ought to be'. It is repeated on Jewish gravestones of the 2oth century as if it evoked some subconscious memory of how Paradise should be depicted, and the explanation that the gravestones have anything to do with the Jewish flag is incorrect for the two are not identical. However, the Jews do have a strong connection with the Star in Armenia, for it was here that the patriarch Noah is supposed to have finally come to rest after being tossed around during the Flood in his ark, which landed on the mountains of Ararat. Note mountains, plural.

The mountains in question are Greater and Lesser Ararat, known as the Mountain of Mars. Whatever the object is on the higher slopes of Ararat, we do not know, but a pictogram or carving at Karada shows an object descending *between* the two peaks, but whether it belonged to Noah or some Martian colonists, we cannot say, if indeed it belonged to either. What we can say is that there may be *two* Arks on Ararat and if anything came from Mars it was *not made of wood*.
There is indeed a place named as Sardar Bulaki between the two Ararat volcanos at an altitude of 1514 ft, some ten times lower than Lesser Ararat. Did the ark land here? If so, which one?

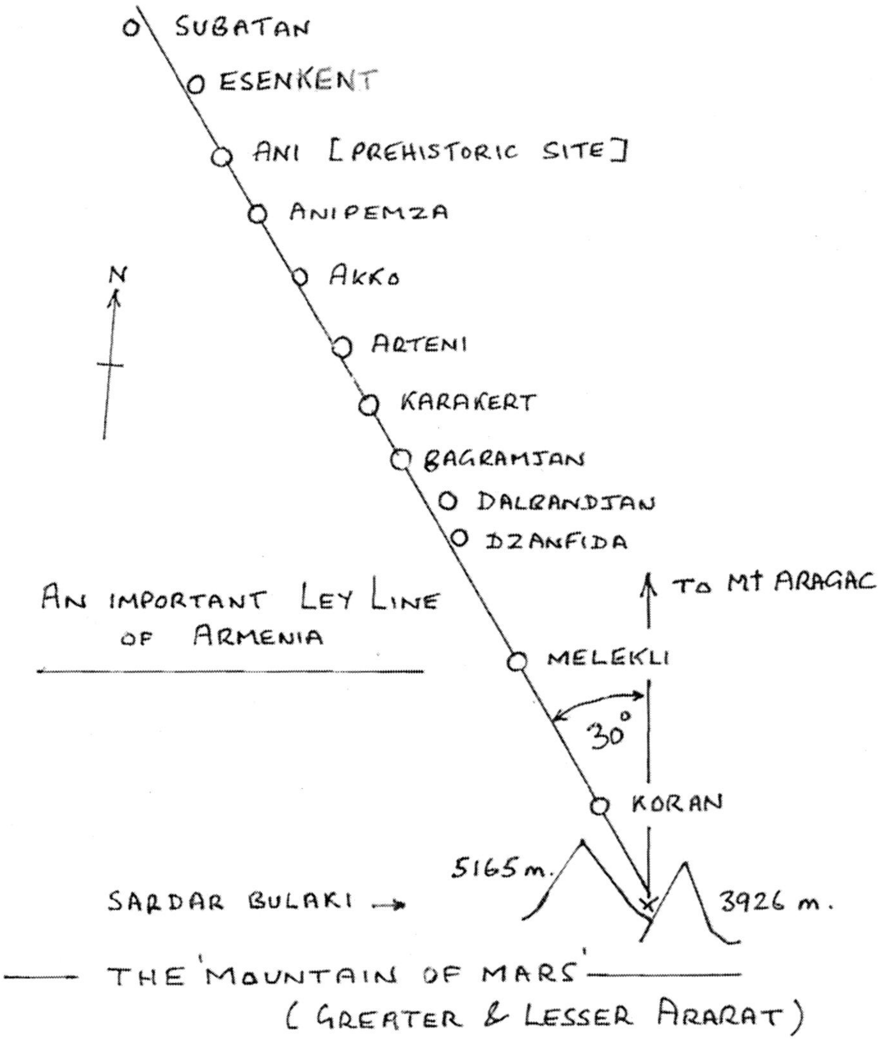
SUBATAN
ESENKENT
ANI [PREHISTORIC SITE]
ANIPEMZA
AKKO
ARTENI
KARAKERT
BAGRAMJAN
DALRANDJAN
DZANFIDA
N
TO Mt ARAGAC
AN IMPORTANT LEY LINE OF ARMENIA
MELEKLI
30°
KORAN
5165 m.
SARDAR BULAKI
3926 m.
THE 'MOUNTAIN OF MARS'
(GREATER & LESSER ARARAT)

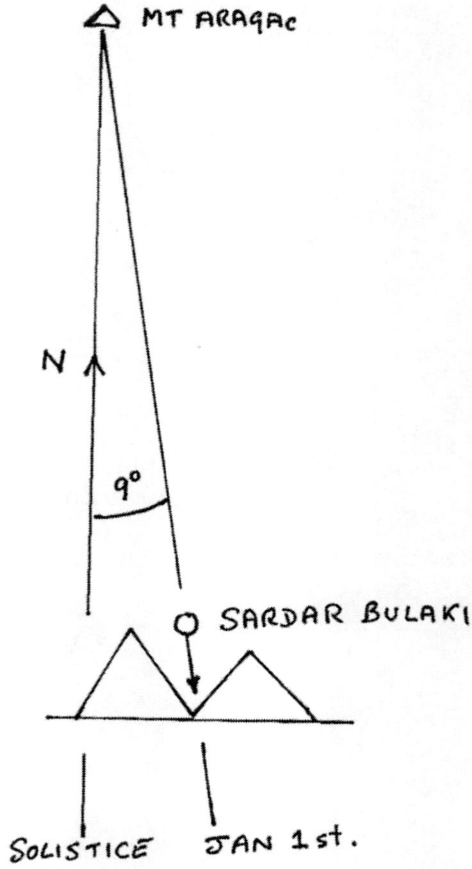

The 9 degree angle is shown between the Solistice and January 1st, matched to locations in Armenia.

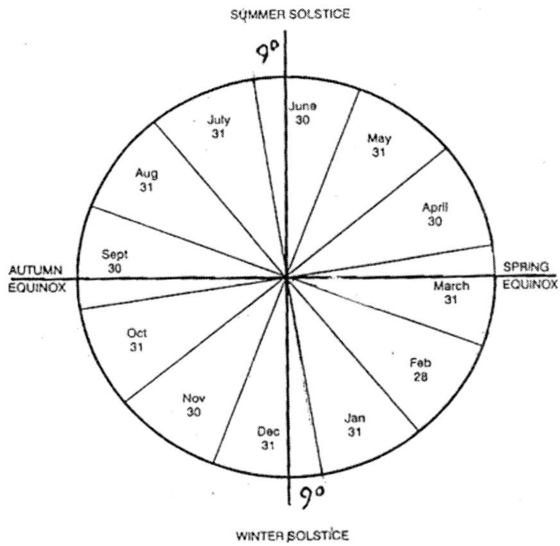

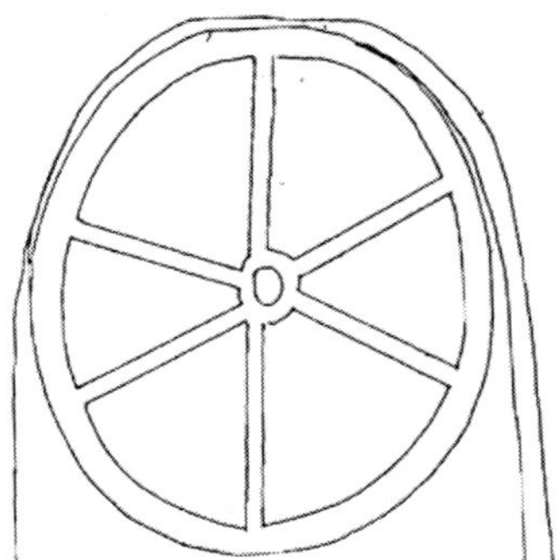

Celtic Cross at St. Iltud's Church

Sardar Bulaki is important for other reasons; of which the first is that a ley-line drawn from here and following the north-west 'side' of the Star passes through or by no less than *twelve* towns, villages, and ancient sites.

The second reason is because of the connection with the calendar we use today, when the year begins on January 1st, and if one should ask why, no answer will be forthcoming. This day is not the shortest day of the year; that will be sometime between the 20th and 22nd of December. But if a drawing is made with with the winter solistice at 180 degrees on a circle, January 1st will be *nine degrees away* at 171 true. Furthermore, roughly *nine* days will have passed between the two days. There will be no need to remind the reader familiar with occult matters of the significance of the number nine, generally connected to evil since the true meaning has been long forgotten.

Now draw a line from Sardar Bulaki to Mt Aragac and measure the angle to true north, which will be found to be *nine degrees*. Coincidence? Maybe.

The most compelling evidence that this tiny unremarkable place has something to do with an ark is that Sardar Bulak(i) can mean *Chief Boatman,* working through about three languages to Hindi via Persian. In 1845, it was described as a Cossack outpost with a well, in the middle of an arc-shaped plain between the two mountains. Many expeditions set out from here when the answers they sought were possibly under their feet.

Fascinating, isn't it.

Although the best maps of Armenia are of Russian provenance* the Russians were not interested in ancient monuments nor standing stones and so they did not include them. Perhaps there are better maps than we have found in Yerevan libraries but one might have to learn Armenian or find a translator before consulting them. We can only suggest that more research is done before Armenia goes up in flames for about the hundredth time. The most important item of equipment would be a SatNav device, and the results should be fed into a computer in order to establish the reality of our findings.

Would anyone like to finance an expedition? If so, write to the Publisher or email.

An unanswered question is what, if anything, is at the center of the Star, which can only be answered by actually going there once a SatNav position is established.

* Russian maps: www.lib.utexas.edu/maps/commonwealth/yerevan_84.jpg

THE MEANING OF EDEN

The Armenian language has boots
of stone that won't wear out.

Osip Mandelstan

THE MEANING OF EDEN

Experts on language, what it is and how it came to exist in the form we find it today, are reluctant to admit that some languages are related despite the existence of hundreds of comparable words in tongues separated by centuries and thousands of miles. Their views on ancient travel by land and sea are formed in some academy and treated as holy writ until proven dead wrong, such as the recent widely promoted theory that there had been 'no white men in America before Columbus'. Really? Not even a few Vikings?

It would be said, for instance, that the Maori '*o te*' meaning '*of the*' is mere coincidence. Perhaps. But to discover the meaning of Eden it is only necessary to consult Anglo-Saxon, which is Germanic in origin, and German has hundreds of correspondences with Persian. Or, one might consult Sumerian, where the word '*din*' existed, meaning a plain. Anglo-Saxon gives us '*dene, dell, dale, den, dent,* which are obviously related, meaning a depression in the ground, but to close the gap we would require books or scrolls that simply do not exist any more, let alone useable dictionaries.

The expert may deny that there are words with the same sound and meaning in Maori, Greek, Minoan and Sanskrit crying "Coincidence!" and denying reality. What a pity.

Unfortunately, it is true that language comparison is not a popular subject anywhere because this is deemed a useless pursuit, while a good percentage of people can no longer adequately speak write or read their mother tongue. But you know that, so we will move on.

This work celebrates the fact that after many years when NASA coloured the Martian sky pink, red, or brown, Holger Isenberg of Germany finally showed the sky to be blue, as it is on Earth, and NASA had practiced deception until 2001 when this book appeared. Every planet with an atmosphere has a blue sky. (The Rayleigh Effect).

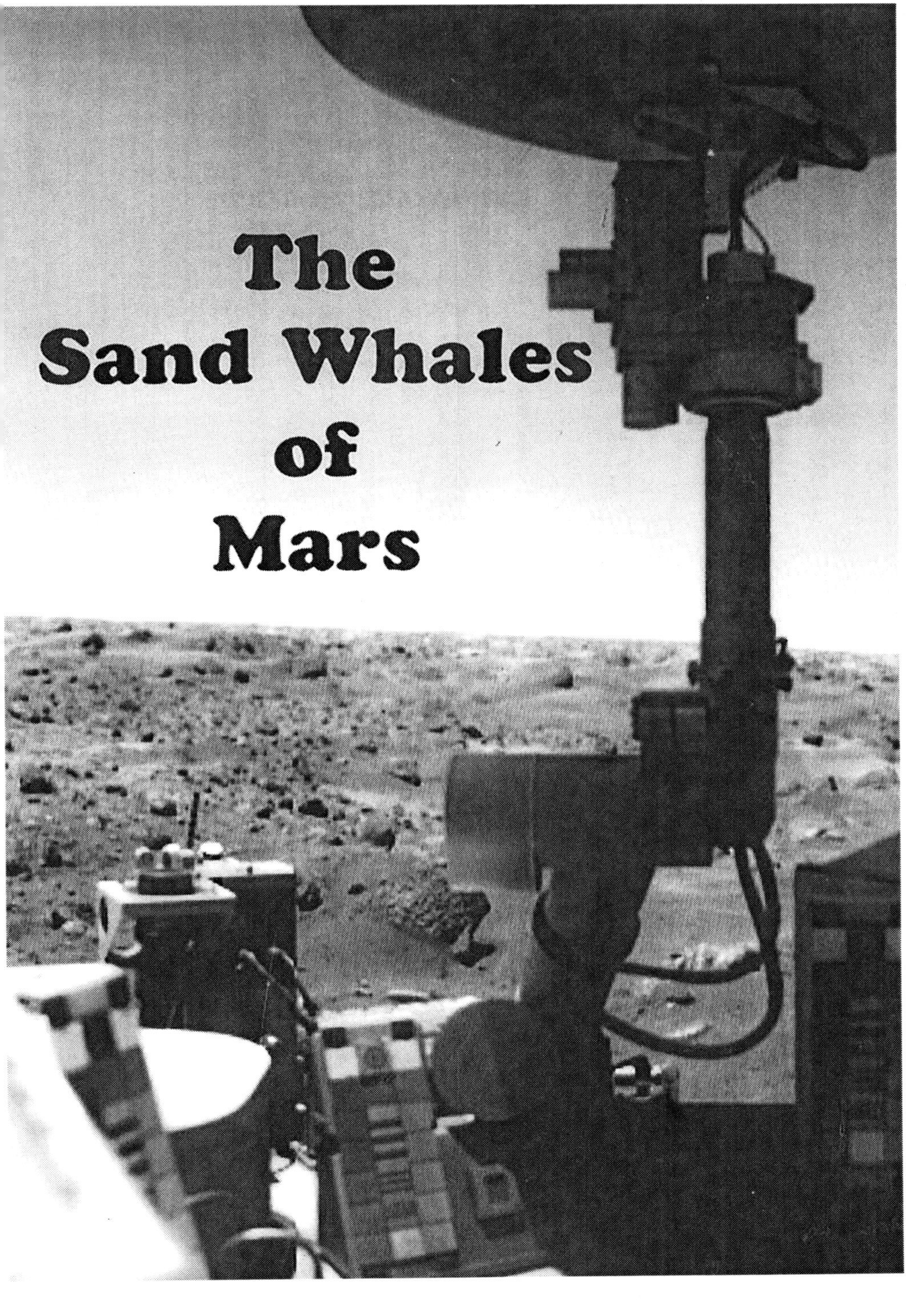
The
Sand Whales
of
Mars

THE VOYAGE TO EARTH

I know the star knowledge
of stars before the Earth –
Whence I was born
How many worlds there are.

Cad Goddeu, trad. Wales.

THE VOYAGE TO EARTH

This chapter contains some psychically derived material which may not be acceptable to all readers, so it is best to put it into italics. The source was Miss Frances Yule, who was casually asked about Mars in 1979 giving the surprising information that life was not extinct on that planet, which is yet to be proved nearly thirty years later. When asked to amplify her account of Mars, she gave the following information:

The ship which came from Mars was shaped like two spheres joined together. It was a one way journey because there was no provision for a return and no landing wheels. Instead, there was a kind of skid. The captain's name could be pronounced 'Orno' in English and there were 24 crew. They landed near Ararat and the ship is still there. If it was examined some interesting things would be found. They also carried seeds and some smaller animals. The journey took 14 months.

Miss Frances Yule does not want to be found, and the most that can be said is that her words were noted down in Ponsonby, Auckland, although she claimed to be an Australian. She also gave many prophecies which were published in South Australia in 1982 as 'The New Life Prophecies of Frances Yule'

This statement was checked, of course, wherupon it was found that the 24 colonists were well known in sacred art and legend, while Orno might have become Onouris, a god of the Greeks. The ship is equally well known, in outline at least, being carved upon a rock at Karada, Armenia, descending between two peaks which are Greater and Lesser Ararat, the Mountain of Mars.

Sacred History Research Expedition

FIRST MESSAGE FROM THE ARARAT REGION

By Col. Alexander A. Koor

Some inscriptions of Karada (Kara-dagh?)

Some inscriptions which I found and copied in the year 1915 in the region of Mt. Ararat:

1.....................................?

2.....................................?

I translated the pictorial signs and they give me an accurate explanation of the origin of the words MAGOG, GOG, etc. (Ezekiel 38:2 and 39:1) This explanation also upholds the truth and accuracy of Chapter 10 and of Genesis. In short, the words Magog, Gog are words of the Sumerian tongue, which shows the ancient origin of the words. Jews incorporated the words through the Assiro-Babylonians. The word MAGOG consists of two words: MA and GOG.

The word MA is a pictorial sign which is  and in cuneiform: 

It means: (1) "origin. (2) "the seeds of the waters (FLOOD)".

The word GOG, his pictorial sign is 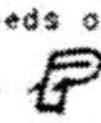 and in cuneiform: 

It means: (1) "the Word of GOD" or "by the Word of God."

(2) "thrown" or "banished by God."

Having two meanings:

Pictorial sign -"originating from (by) the Word of God" and "the seeds of the waters (FLOOD), thrown by GOD," are both amazingly coincidental with BIBLE texts. Noah came in Mt. Ararat's region with the Flood, that is, with the waters. Likewise, the first meaning: "Originating with the Word of God", is upheld in Chapter 1, Verses 1, 2, 3, The Gospel of St. John.

The word GOG is met with in the text "Ishtar", from the library of the temple Ishtar in Nipur, and also in the text of "7 tablets of the Creation of the World", part 6, p. 11. For linguistical disbelievers of Ancient Ball "Sumerians", pictorial signs, as proof.

Pictorial sign  is read : TO KOOR and means: to come to rest on mountains" or "peak". Cuneiform  KOOR" means, mountain. (Koora or Kura.)

Pictorial sign  is read: TOOB and means: "to filled earth." (TUB-AL).

Cuneiform: 

Pictorial sign  is read E-GA and means : "the flood, night descended from upon."

The signs: 

they may be translated as follows: "These thrown by God, by waters came to rest on the mountains when the night and the waters (flood) descended from upon the earth......."

Since these four pictorial signs alone, through the centuries, were well preserved from two or more lines of damaged inscriptions (possibly by earthquakes), I copied only the four signs. The place where these signs are inscribed is called "Karada". (m. b. Kara-dagh).

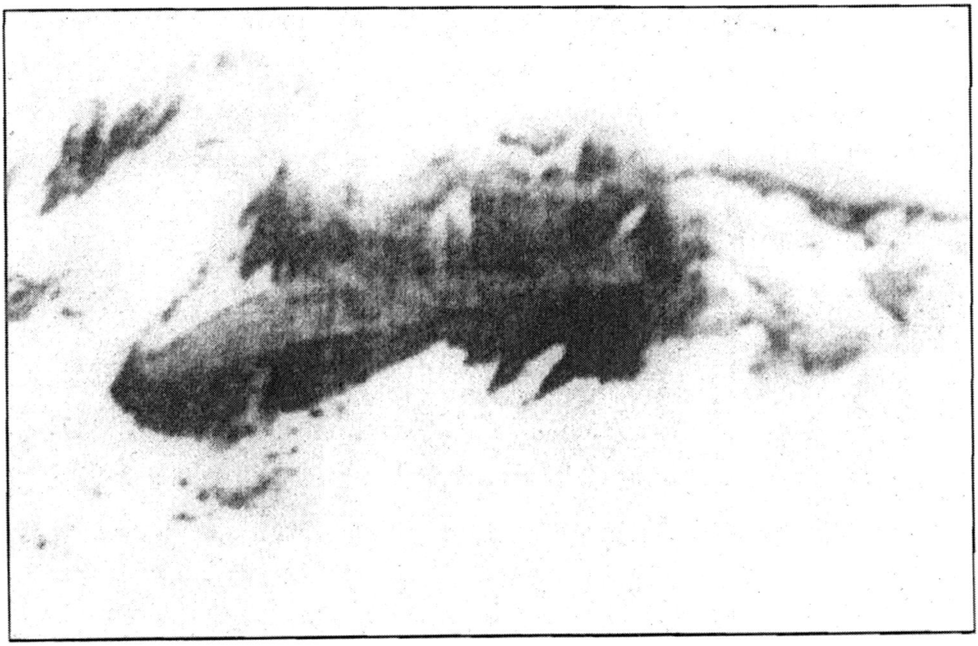

Photo taken by the USAF. text reads:

(Illegible)...HQ AR. 17 June 1949 14000 ft 24°
(coordinates illegible).. Mt. Ararat.

Photo enlarged and enhanced by Paul J McLeod, 2001.
From 'The Sand Whales of Mars' pp. 39, 40.
Black Rabbit Press 2001.

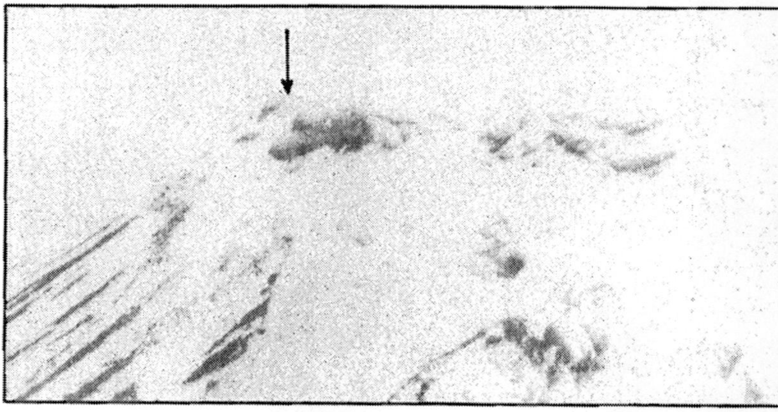

The Mountain of Mars

The Mountain of Mars is well known in symbology, and is illustrated in Cirlot's 'Dictionary of Symbolism' (See appendix) in the manner shown here. The descending object can be replaced with various other symbols such as a short straight line, the Fleur de Lys, or a crown, which might give pause for thought in that the source of kingship is not Jesus Christ, the Merovingians, nor any or all of the French kings, but Mars.

Viewed from certain directions, the two great pyramids of Giza repeat the construction, while any twin peaks or notch on the skyline might be incorporated into a prehistoric version of the Mountain of Mars. It is the home of all things double, and the reader may recall that the number of Mars is two, as in deuce or Dyaus, an early Greek name for God. This became Deus.
Although the exterior of this mountain was held to be uninhabited, it was thought to be hollow, the dwelling place of some hero or deity who lies asleep awaiting a call to restore his realm as it was in days of old. In fact, Ararat has seldom been climbed, let alone inhabited, except for a monastery at its foot to the north which was destroyed not so long ago in a volcanic eruption. As for the Martian Ark, there are pictures of what might be a ship on its snow covered slopes, but the Turkish authorities, being Moslem, do not care to have Ararat investigated considering it is in Turkey near the Armenian border and hostilities between Armenia and Turkey could be said to be ongoing, if only as a war of words. Pictures of this object were reprinted by Paul McLeod on both the Net and in his booklet 'The Sand Whales of Mars' available from this author, see page 38). Many other works discuss possible ships or relics on or near Ararat, but it is germane to state here that Welsh legend gives its name as the 'Kyd', and the builder as Menwyd. The Kyd (kid) is the offspring of the ram, which is a martian animal.

Left, with sling, David. Right, with sword, Goliath.
Lower: six and twelve lobed florettes.
Below: a sleeping Ram. Above: the Constellations.

Church in Armenia.

The ram is quite surprisingly known to be red, which recalls the colour of Mars, and this is mentioned in Genesis. To link the ram and the Star of the Garden of Eden is quite easy, for an Armenian church has a carving of David and Goliath, beneath which we find a sleeping ram and a carved circle, within which is the classic aster or six pointed flower that equals the six pointed star. We are not fooled! Above, there is a script written in Armenian, of course, which someone will have to decipher although symbols can reveal just as much to the initiated. There is a second carving which might be a twelve petalled flower. The sleeping ram is Mars as it is today, awaiting a new Spring that could be engineered by what is known as 'terra-forming' if we had the money, time, and will.

Here we note that the actual physical colours of Mars, the red sands and the white polar caps, are repeated in the dress of many hero figures, from Saint George (who is reputed to have been án Armenian!) to Horus of ancient Egypt, the Nordic Odin or Othinn, crusaders, the Red Cross organisation, and several characters depicted on Tarot cards, especially some by the noted Golden Dawn occultist, A.E. Waite.

It is not known what colour the Martian ark was painted, but it carried many tons of grain on board, not for eating but as a radiation shield similar to the carbon used in nuclear reactors and employed for the same purpose- to stop radiation from penetrating where it is not wanted. Space ship designers take note! The design of two connected spheres, was used because they could rotate about a central point, thus providing some artificial gravity, which is now known to be essential on long space voyages if the body is not to deteriorate.

STIKI, VISHAPS, AND MARBRAS

According to the Coral Coast Art Gallery, dead men in Armenia would and could become dragons.

www.theserenedragon.net/Tales/armenia

STIKI, VISHAPS, AND MARBRAS

There are supposed to have been several different horrific monsters living in the Caucases from Armenia to Albania, of which we can mention Bolla or Bollars, vicious Kulshedras, some Odz Manouk, the dreadful Stiki, a flappathon of Svara, the ubiquitous Vahagn, and the multi-tonne Vishap. Were any of these names bestowed by Martians, did they bring any of them here as pets, breed them from some spare DNA, or are these names simply invention?

Most of them are fire breathing dragons with a penchant for consuming humans, of whom young virgins would be the most desirable dish on a dragon's menu, and for more info you are directed to www.theserenedragon.net where a useful dragon chart is displayed together with notes on sources, types, origins, and tales, which you will find wildly enthralling if you are into the science of dragonology.

To save any puzzlement concerning these scaly monsters, the prototype was seen in the heavens at the second annihilation of the majority of sinful mankind, which happened in 3237 BC and is still remembered around the autumn equinox at Halloween by compulsive behaviour such as lighting blue touch papers and retiring immediately, burning guys and witches, and frightening innocent cats, dogs, and old ladies. At that time, over 5,000 years ago, a 'dragon' passed over terrified humanity who thought the 'End' was nigh and just as it so happened, it was.

The results are well known, being Noah's Flood, as you can read in the Bible and in dozens of other accounts from around the world. It rained cats and dogs because the sky monster always brings millions of tons of dust which cools the climate and produces continuous rain.

The earliest Višhap (Vishap) stele can be dated to between the 18th and 16th centuries BC, created in pre-Urartian times according to an 8th C. BC Urartian inscription.

www.projectdiscovery.net/journal/vishap_stones

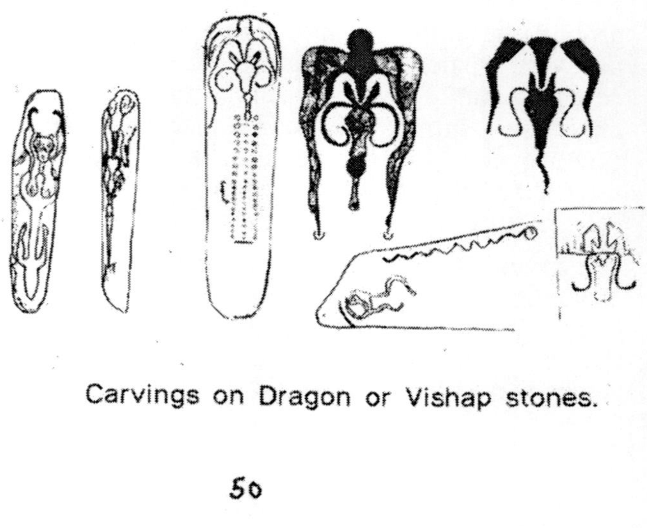

Carvings on Dragon or Vishap stones.

There are Vishap stones on Mars.
This one protrudes from the floor of Candor Chasma.

After many years painful research, it became clear that the next visit of the Dragon (or Odz Manouk or whatever) is due in 2115 and it will sweep past the Earth twice within six months. There is a good description of the results, published by 'Nexus' magazine, supposedly written by Mother Shipton. You may say - What is that to do with me because I won't be there- but your great-grand-children might witness the arrival of this object if they are born after 2030 or so. In fact, you may well suffer the results of increasing sunspots to a maximum never before witnessed in your lifetime, which will affect communications. There are any number of books on the extra member of the Sun's family which you can obtain through a public library, notably those by Sitchin and Velikovsky, as well as our own more modest efforts published in 2000 and 2003.

It may be called the 10th or 12th planet and has several names apart from the well known 'Dragon' as you will discover. The Dragon has an 892 year retrograde orbit.

Humanity set up memorials to this repeating event so that people would never forget, which they did, of course, thus necessitating more memorials after the following visit, 892 years later. That is why Armenia is covered in stones which are an impediment to navigation of the highways and byways, agriculture, and forestry but are nevertheless handy places to rest one's rear end. Many of the lumps of rock were gathered up and placed in or near churchyards where they remain. The largest are called Vishaps, being 10 to 20 feet long and shaped rather like a cigar. Armenians who have studied them conclude that the message the ancients wished us to learn is that disaster can be avoided if one conducts a bull sacrifice on or near the Vishap, when the 'serpent' or dragon will be induced to go somewhere else and kill 95% of the unlucky populace in that unfortunate locality. If one wishes to personalise this age-old routine, then take

Saint George, who was supposed to have been an Armenian, and let him slay however many dragons he can find. We ought to mention that the forthcoming dragon spectacular blockbuster event of 2115 is described in what may be called the 'Nexus' prophecies of around 1995, which are all over the net and attributed to Old Mother Shipton of course, who must be performing a considerable rate of rotation in her grave, God bless her, because she wasn't responsible.

At this point in the narrative, if that is what it is, we must consider the Marbra as divulged in our chapter heading, which is a modern invented name for a creature seen on Mars- discovered a few years ago but not accepted as a living creature by anyone important. Of course, the disbelievers in Martian animals have no choice but to say any strange object is a rock, a crater, or a 'transient phenomenon' (whatever that is but it covers a lot, doesn't it.) Well we will show you marbras. The name comes from 'Mars' and 'Zebra' because of the patterning on the creature's back, which any Areologist (knower of Mars) of the head-in-the-sand fraternity will say is exactly that- ie wind-blown sand patterns or ripples. Having their heads stuck firmly in the sand, they should know.

There are pictures of large martian animals on page 57 of this book, of which we have repeated the Marbra three times at different magnifications. giving the source image number. It is found roughly 2/3 down the strip to the right, and was discovered in August 2003 and published on **www.martianmania** and **enigmasmania**. These things are larger than a dinosaur but probably subsists on the creatures living in the soil much as a whale consumes tiny krill.

The things at the lower left of our picture are called 'sand-whales' which might serve as illustrations to the well known fictional work 'Dune'.

EDIACARA FOSSILS,
SOUTH AUSTRALIA

600 Ediacara fossils have been found of many varieties. They existed before there was any oxygen on Earth and are unclassifiable by science. They could live on Mars as it is today.

Picture courtesy of Paul McLeod, from 'Sand Whales of Mars'.

It is possible that a Marbra is nothing more or less than a huge gasbag running on hydrogen which reduces its weight, even on Mars where the gravity is 38% of that of Earth. It may obtain the hydrogen by splitting water which would also give it oxygen, assuming that it needs any. This apparently ridiculous statement makes sense when it is realised that there were once creatures on Earth called Ediacara, found for instance as fossils in South Australia, that managed very well when there was no oxygen in Earth's atmosphere, many millions of years ago. Alternatively, there are organisms living near deep sea volcanic vents that use sulphur for energy. However much or however often these facts are pointed out to Mars researchers, they act as if they are permanently blinkered or perhaps following a 'line' given them by some unstated and invisible superior authority.

Here, then, we will turn to Mr Phillips who was invited to write his thoughts for you on these matters -

All good and real science begins with an open mind. I must confess that when I began the investigation of life on Mars I was biased against finding life. There was no atmosphere, right? The photos are taken from a space craft a couple of hundred miles up with cameras that can only resolve a couple of meters/pixel. Then there is the general prejudice against such a discovery in the scientific community at large. They are still trying to collect more data to evaluate relativity and quantum mechanics. Telescopes have always been valid tools for other scientific disciplines as are microscopes. High speed cameras are used to capture sub-atomic particles on film. Particles that can't be put in test tubes or under microscopes.
Photographs have always been valid for other

scientific disciplines for analysis, theories, or validation of theories. So, if one was to be scientifically objective about life on Mars or elsewhere then a photograph showing a large life-form should be considered evidence, especially if it is similar enough to known life-forms. Am I missing something here?
What kind of analysis can be done on such a photograph? The symmetry of the features should be apparent. That is, two eyes that are separated by a bridge-like nose and positioned above a slit or mouth. Well, we have such a photo from Mars. In fact, we have two of the creatures taken a few weeks apart. There are many examples of creatures on Earth that remain stationary or even dormant for extended periods of time- sometimes months or even years. The fact that the creature appears virtually in the same pose at two different dates is not a surprise. After a snake eats it can remain stationary for 30 or 40 days. The eyes do appear open in one photo and squinting in another. I have had others tell me they think they can see the pupil when they blow it up 4 times normal size. I saw the same thing when I finally printed it on photographic quality paper.

www.msss.com/moc_gallery/e13_e18/images/E13/ E1300965.html
www.msss.com/moc_gallery/e13_e18/images/E14/ E1401522.html

We are going to have to rethink our theories about life. Before these photos were taken there was no reason to ever expect more than lichen-like vegetation or fungus. If the evidence I present here becomes accepted, then scientists will fumble for (cont)

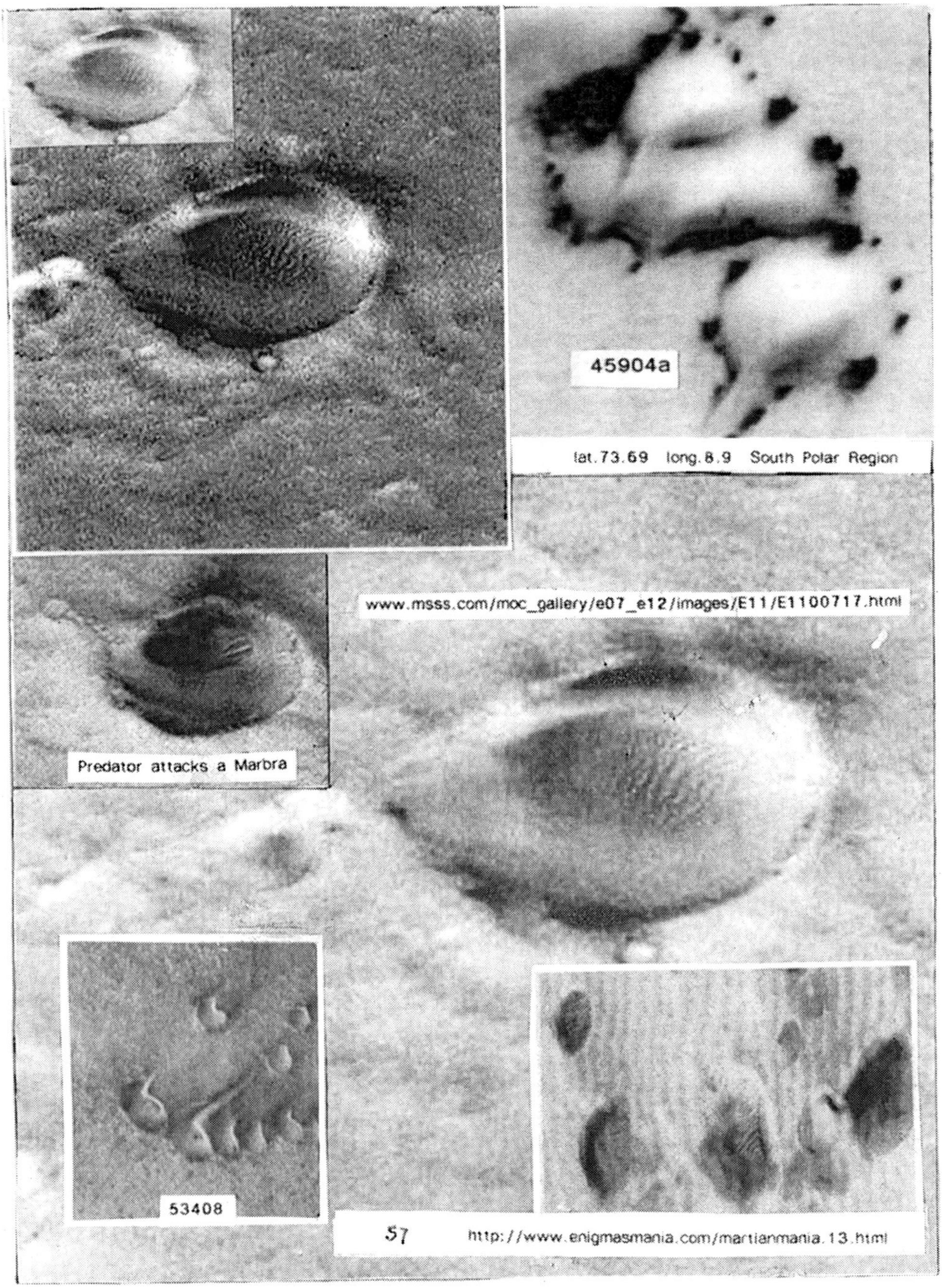

45904a

lat.73.69 long.8.9 South Polar Region

www.msss.com/moc_gallery/e07_e12/images/E11/E1100717.html

Predator attacks a Marbra

53408

http://www.enigmasmania.com/martianmania.13.html

explanations while their shock wears off. I have to sympathise with anyone who has stuck their neck out in the past and made unsubstantiated pronouncements about anything as the absolute truth. I would just say- wipe the egg off your faces and start helping to figure out what this might mean and what the bigger picture is. Does life permeate the Universe? Wherever there is rock and soil a symbiotic relationship starts with living organisms that cycle the material from living to dead and back again. Will a new paradigm and the younger scientists develop a new philosophy that is the opposite of the old? Will they look at reality with wonder and amazement and ask questions rather than dismissing thought?

The most compelling evidence I have for the existence of life on Mars is near the bottom of the wide-angle photos in Part XIV of my website.

Mr Phillips presents two rather small pictures of what seems to be a live animal which moves between shots. If it moves, he says, it must be alive. Mr Phillips continues...

The desert plains of Mars have blotches that change with the seasons. This could be water seepage, while there are mounds that could be fungus or some other kind of vegetation. Many astronomers writing before the 1960s wrote book after book about changes on Mars; some imagined that they could also see canals, but now we have cameras up there and can see large creatures that might live off the vegetation.

We heard for decades that there could be no water or significant atmosphere on Mars, but now we hear periodically about more and more water on Mars, and also the Moon, of all places. Does anyone see a pattern here yet? I can't wait for the atmosphere to start thickening and the amount of UV radiation to start decreasing with each new report!

Mr Phillips refers obliquely to the 'boulders that get up and walk away', as some wit said at the beginning of this century, which are probably huge fungi that grow and fruit in season, while lying dormant and nearly invisible in the Martian winter. Please use your computer to bring up Mars Orbital Camera images **MOOO2906** and **M0202258** which were taken at an interval of two months. Mr Phillips says -

Mars is a very curious place. It hides things in plain sight. It is easy for even the trained scientist to glance at a photo, see something familiar, and then neglect to examine the scene more closely, such as might be done by comparing images of the same site taken at different times. That is where 'anomaly hunters' come in handy helping to keep the astronomers honest. I have tried not to give them quick and easy 'at a glance' anomalies to airily dismiss, as they do, all too often.
Another thought to ponder is this- when something is blurry while the rest of the scene is sharp, could this be because the something was moving when the image was taken?

I have listed several criteria for identifying and verifying the existence of life by photo analysis -

1 *Did the scenery change in subsequent photos?*
2 *Do the images show life-forms similar to Earth creatures?*
3 *Does the face possess symmetry?*
4 *Is there blurring which indicates motion?*

Emails welcome...

keith_phillips@comcast.net

Keith Phillips has complained that his website has suffered attacks in the past emanating from Moscow, such as the occasion when his site was temporarily replaced with inferior Spanish pornography. You may not believe this but it is true. Further, NASA copied the name of his website so that a multiplicity of similar sounding names overwhelmed the private site, which lost position and rankings in a most peculiar manner, until www.mars-mania was forced to transfer elsewhere, to www.enigmasmania.

One wonders if it is possible to 'demote' an unwanted website for most searches do not go beyond the first or at most the second page. It is absolutely certain that someone or some committee decided a long time ago that there would be no kudos nor applause given to any government paid or assisted researcher whether they worked for NASA or any other organisation who did not toe the official line. It may be that they will consider if there was once life on Mars, although even that is considered an open question, but it is taboo to openly consider the possibility that life may still currently exist there. The most that they can think of is that maybe we, the all powerful lords of Creation, could change the ecosphere of Mars to something more to our liking.

Anyone with a mind of their own can find a pack of anomalies of Mars on many websites devoted to the subject, including forests of flattened lichen-like trees bigger than you might imagine living near the South Pole, some very large animals of which we have shown you a selection, plus the fungus which probably comes in several varieties. Then we have skeletons, shells, molluscs, lichens, indescribable things with no names, and gases in the atmosphere that would not be there were there no life, such as methane and hydrogen. Even the nitrogen in the Martian atmosphere proves life because it comes from earthworms, believe it or not, and that is where the nitrogen on Earth came from.

There are a very few people who actively look for anomalies on Mars; an activity which if pursued for too long tends to reduce the seeker to tears or compulsive behaviour; often ending in divorce. Pictures of anything found usually end up on the Net but may have a limited lifespan, determined by the amount of attention the newly discovered anomaly attracts, and we have ceased paying attention to faces on Mars simply because they don't actually do anything and soon there is no more that can be said about them.

The public soon lost interest in viewing astronauts kicking up dust on the Moon and performing stunts. Perhaps NASA was not inventive enough and so the mere sight of a couple of clowns clumsily performing tricks on the Moon's sterile surface became boring and a five minute wonder. More was required of NASA but it appears to be an organisation solely devoted to its own survival and not genuine scientific discovery, so that the obviously interesting places to land were bypassed in favour of the Martian equivalent of the Atacama Desert. Had probes been sent to the vicinity of the South Pole of Mars, where the majority of the martian life exists, and deployed intelligently, we would now have clear unambiguous pictures of martian life in all its varieties instead of repetitious pictures of nothingness sent back by a couple of solar powered motorised wheelbarrows sent to the middle of nowhere.

On one occasion, the surface appeared to be mud, but wait! There couldn't be any mud on Mars so it wasn't mud. Perhaps it was some other chemical but we'll never know. We aren't meant to know. Similarly, there are plenty of pictures of 'glass tubes' or 'worms' that interested the likes of Arthur C. Clarke, but even he was unable to persuade NASA to take a closer look at these wonders of nature, probably constructed by worm-like creatures because they need a habitat. If you wish to denigrate something, don't give it a name, or call it something inappropriate like 'a glass tube' when everyone knows there are no glass factories on Mars. The worst problem with whatever is found on Mars is that it is 'too big', but that is only in comparison with what is 'normal' on Earth. While evolution proceeds in the direction of 'faster, smaller, lighter, brighter' on Earth, Mars life needs to be 'fewer, bigger, slower' because resources are scarcer, including oxygen where the only available supply might be in the form of ice.

If NASA lost its way about thirty years ago, refusing to answer the simple questions asked by the very people who pay its bills- the American Public, then the task of making sense of new worlds became the province of the amateur astronomical fraternity; derided, under-funded, ignored, and vilified by turns.

What were these questions that NASA would not answer? Simply put: Are we alone in the Solar System/Galaxy/Universe? Is there life on other planets and if so, what does it look like and what is its metabolism? Making a start on answering these questions demands honesty in the first place; sufficient to enable frightened scientists to admit that there are things 'out there' on the surface of Mars that nobody really understands, and one must start by giving the things names.

We, the authors, conducted a search for martian life forms starting over seven years ago, employing the premise that since camera resolution was so bad, all we could look for were monsters. Nobody really expected to find any but we were going to look hard. In August 2003, a thing we named a 'Marbra' was found and soon there were a few more on display as we learned to recognise what we were looking at from odd angles and viewpoints dictated by the camera. There was no way we could prove any of these martian heavyweights were alive and not mere rocks or holes in the ground, except for the obvious- if it moves, it lives.

So everybody sat on the proverbial fence which is quite difficult with one's head embedded in the sand, until four years had gone by and the Chief of MSSS thought it was probably safe to take more pictures of the supposed monsters, hoping that the things had not moved an inch in any direction. How wrong can you be. Every single martian monster we had ever discovered had taken off at whatever lumbering speed it was capable of in whatever direction it chose for whatever reason that had occurred to its probably minute brain! Thankyou Mr Malin.

Here we would like to remind you to take a look at the Marbra which has something unusual on its back that might be a predator. Blow it up and you will see a spherical dark creature with two rows of 'knobs' along its back, probably munching away at its prey, or it may be a parasite or even the male of the species. Note that Marbras leave a trail of something that would have to be called excrement, resembling cow-pats.

Now Mr Malin of MSSS did not take pictures at random for nearly every one of his shots was straight down the middle of the martian bowling alley, on target, and clear as a martian sol. He knew exactly what he was doing but we don't know why he did it. People have given up asking anything where NASA is concerned, which is a pity when so much needs to be investigated, but it can only be hoped that enough people will read this book to make a difference.

AREOGRAPHY

being the geography of Mars

Our decision is that there is no
life on Mars, there never was,
and we ain't gonna talk about it

NASA doublespeakperson

THE GEOGRAPHY OF MARS

The reader's attention is directed to the old Persian map of the world, complete with its depiction of Paradise as being at the center and the four directions which correspond to the solistices. This is taken from Zoroastrian mythology and was preserved by the Sufis who are a Moslem sect. Here, Paradise is called the *Airyana Vaejah*, or *Eran Vej* in Sufi terminology, while the directions north and south are also given in order to divide the world into six *keshvars*, or regions, which were surprisingly thought to be uninhabited!

The small circle of Paradise, the *Airyana Vaejah*, is described as the 'first and best dwelling place', while the larger circle surrounding it is called *Khvaniratha*. There are several alternatives to this name such as the psychically derived *Karraveth* (Frances Yule) and the Mormon *Kae-e-vanrash* (Joseph Smith), while certain persons who wished to disguise this name substituted an H or an X at the beginning of the name, making *Xvaniratha* and *Hvaniratha*. It may be the same as Aleister Crowley's *Hadith* and the *Khunrath* of Michael 'Khunrath', a mediaeval occultist and alchemist. It is the same as *Kadath* which appears in the Necronomicon, allegedly given by one Abdul al Hazred to the Elizabethans Dee and Kelly in the following words:

"Beware, O Man, beware, of those with mitred heads who tread in darkness the ramparts of unknown Kadath"

The Necronomicon is a collection of sigils (signatures of demons) and short passages which should be approached with caution.

THE SEVEN KESHVARS

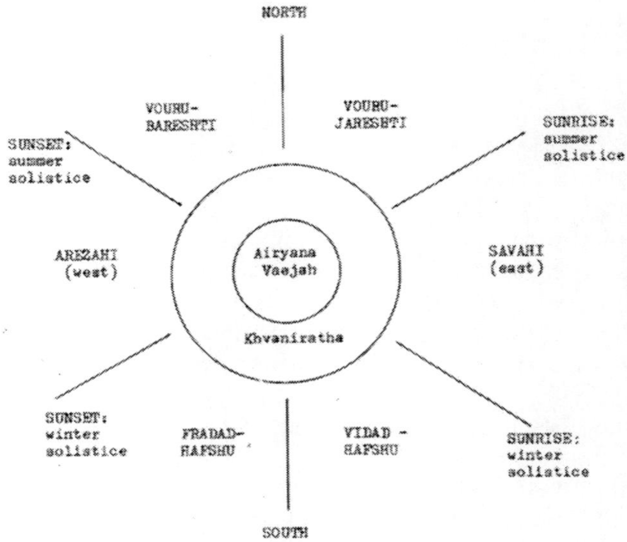

After Sufi and Zoroastrian texts: showing the globe, Khvaniratha, the Garden of Eden as Airyana Vaejah, and the six directions which with the Airyana Vaejah, make seven Keshvars. This diagram only works in the Northern Hemisphere.

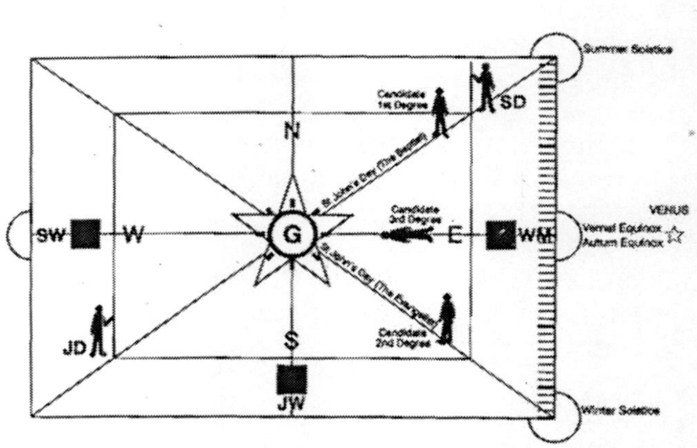

Ideal layout of a Masonic Temple.
The divisions to the right reflect
the 23.45 degree tilt of the Earth.

This text forms the basis of the Cthulhu mythos as exploited in the writings of H.P. Lovecraft whose imagination was stirred by mention in the Necronomicon of unknown creatures who might just as well have lived on Mars, for they were certainly not of Earth, such as the *Yog Sothoth* if memory serves.

Now if Khvaniratha is a globe, and we can assume that not everybody thought a globe was flat, it might just be that the concept and name was not Terran but originated on Mars and is the martian name for that planet, which cannot be proved. It can be said, however, that the early inhabitants of Paradise were given a set of names that could just as well apply to Earth as to Mars, as if there had been an effort to retain a memory of Mars by giving martian names to Terran features. We did the same when we discovered new lands such as America and Australia, bestowing old names like York and Perth on places that had no real correspondence with towns in the 'old world'. This came unstuck sometimes when the colonists from Mars told the locals of features of Mars, such as a polar island, which could not be correlated with anything on Earth because no such island exists.

We will move on, then, to an examination of survivals of lore about Mars contained in what are termed *Yashts*, or prayers, preserved in the Zoroastrian religion.

Armenian lore expanded to other nearby countries, firstly to the south-east because communication was facilitated by major rivers, dictated by the ancient advice that one should enter Armenia by the mountain passes and exit by the rivers, which was the fastest way out in case one had not been welcomed with open arms! When in the mountains, exit by the river system because walking down spurs will inevitably result in one facing an impassable junction of two rivers.

Proceeding south-east from Armenia, one comes to the land of Mesopotamia, otherwise Iran and Iraq, where the Zoroastrian religion flourished, promoted by one Spitama Zarathustra. Unlike the Armenians who had no system of writing and may even have held writing to be dangerous to security, the peoples to the south-east had writing (cuneiform) so they could and did write everything imaginable on clay tablets. Much of their writings have survived to this day, particularly when baked in a fire- deliberately or not.

Among the survivals, we find the *Yashts* or prayers, seemingly in the original form of many thousands of years ago, which indicate how one should pray. Worshippers could entreat their deities to grant a petition or request, which might or might not be granted. Curiously, the principal deity was not Ahura Mazda of Zoroastrianism, but a goddess known as *Ardvi Sura Anahita* or *Ardvin Asura Anahit*. An ardvin was simply a spring of fresh pure water, found at the headwaters of a mountain stream, while this triple barrelled name has survived as simply *Anahit*. The places where one might conduct the appropriate ceremonial are marked, even today, by huge long boulders called *Vishaps* which are carved into the shape of a creature having a form between a whale and a squid, which could never have been seen in Armenia, many miles from the sea.. Although probably once upright, many vishaps are now fallen and bear motifs including the swastika or the cross or the stylised ram's head of Mars. Perhaps vishaps lived on Mars- many many years ago but the stones are mute.

An alternative site for conducting the prayer ceremonial was on the banks of a river such as the *Daitya*, now the Kűr river, which has the Aras river as a major tributary. The major river system on Mars is called the Valle Marineris today, which flows from the volcano Pavonis, as we call it today, once *Hukairya* in martian, while the major martian river flowing east was called the Daitya.

Hukairya is the central peak of the three equatorial martian volcanos; the other two being The Mountain of the Dawns to the north-east and The Peak of Judgement to the south-west. Nearby there stands the highest volcano in the Solar System, so far as we know, called *Har* or *Hara*, which is Olympus Mons; from which is derived the name *Horus*, otherwise *Heru* or the Maori *Hori*, among other like terms, which the reader may collect at leisure for the word is universal. None of these features of Mars were seen by earthlings until the invention of the telescope, and certainly not in detail until the late 20th century. It is only now that the truth of Zoroastrian Yashts is apparent... but there is more!

Water was held to be sacred on Mars, as it was on Earth by Aryans, who seemed to worship only two things- fire as for instance the Sun, and pure water. This still features in Terran religion, especially among the vanishing Mandaeans of the area, who outwardly practice Islam as a survival technique. How on Earth immersion of people in cold water washes away sins I have no idea, especially babies who are mercifully only splashed with water and cannot have any sins whatsoever. Originally, the headwaters of certain rivers were sacred as they were fed by snows. In the ancient Aryan speech, these sources were termed *ardvi*, meaning lofty, while an *ardvin* was an infant stream.

A clear picture of two intelligently designed and constructed objects on **MARS**, obviously used to handle waste liquids, probably originating in a mining operation, consisting of two containers with ancilliary pipework discharging into a black waste pond at the upper left. The L.H. container may be connected to a sluice arrangement for the recovery of solids while the R.H. sphere has a high pressure pipeline and discharge nozzle, that on Earth would be called a 'monitor', attached to a rectangular base. Note the hold-down blocks showing the system is under pressure. From Mars Image MO307150 which someone obviously forgot to scrub out. Oh dear.

Picture courtesy of Paul McLeod, from 'The Sand Whales of Mars'. 2001.

Modern research has shown that there was only one ocean on Mars, which covered the northern hemisphere, since this is lower in altitude than the southern hemisphere. The Zoroastrians, the inheritors of Aryan traditions, believed there was a single world-girdling ocean named the *Vourukasha Sea*, which is true of Earth as it was of Mars until their ocean dried up, firstly to a 'sea of reeds' and then completely.

As we read in the *Tir Yasht* or prayer in the holy book called the *Zend-Avesta*, the Vourukasha Sea was notably rough–

31 *He (Tistrya) makes the sea boil up and down; he makes the sea stream this way and that; all the shores of the Vourukasha Sea are boiling over; all the middle of it is boiling over.*

Boiling over?? Is this any ocean of Earth? Hardly. Waves, yes, boiling over? No. Why should the martian ocean boil? The answer lies in lower atmospheric vapour pressure, lower gravity, and high speed winds. Then we learn of the single island that stands in the Vourukasha Sea –

32 ...*and vapours rise above Mount Us-Hindu (or Ush-Hendeva) that stands in the middle of the Vourukasha Sea.*

The Vourukasha sea was, by definition supplied by the term '*vouru*' meaning north, in the north of whichever planet is under consideration, and so far as Earth is concerned, the northern sea is the Arctic Ocean and it has no central island. In distinction, the one-time martian ocean had a *single* rather low-lying island at the *North Pole* of Mars. Now you can try to persuade cartographers of Mars to give it its name back.

THE VOURUKASHA SEA

NORTHERN HEMISPHERE, MARS.

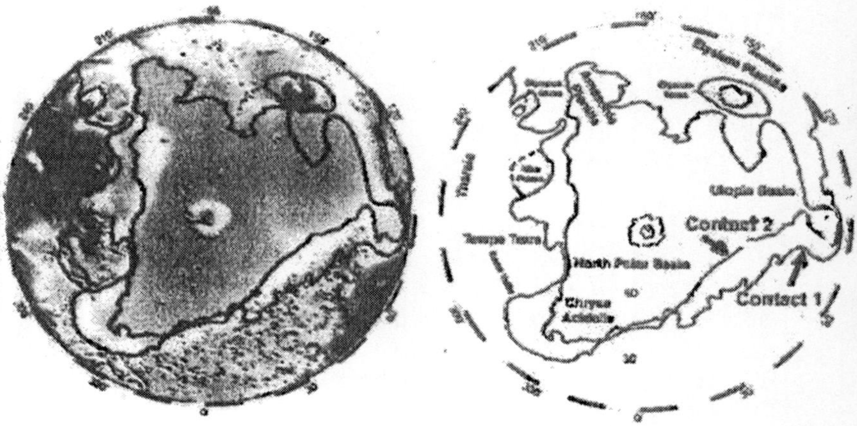

The existence of a one-time Martian ocean was denied by Malin and Edgett of MSSS in America, which oversees the production of Mars Photos for NASA, but laser altimetric surveys of Mars taken from orbit established that there had once been two shorelines here shown as Contacts 1 and 2. MSSS then retracted its earlier statement that no ocean had existed. Messrs Crowley and Pollock had described that ocean in 'Return to Mars' ten years earlier. Note the central island described and named in the Zoroastrian texts dating from several thousand years ago.

Original maps of the northern hemisphere of Mars by NASA. Reproduced from 'The Sand Whales of Mars', BRP, 2001.

The Vourukasha Sea is said to be salt (Tir Yasht 8) while winds blow powerfully all around.. They flow over all the *Keshvars* (divisions of the circle, also *Karshvares)* of which only the central circular *Airyana Vaejah* (Paradise or the Garden of Eden) is inhabited, they say, by mankind. The other six Keshvars are purely directional and so amount to nothing more than an Aryan compass rose. This island called *Ush-Hendeva* may also be called *Ausindom* (Bundahisn XII:6) and was said to be a giant ruby; said elsewhere to receive its waters from Mount Hukairya, which we have dealt with previously. Many legends around the world say that there is a giant jewel at the North Pole made of some gemstone like sapphire or ruby. Maybe the martian polar island would appear red if all the snow melted...

To return to the yashts, it is said that the holy water which was the well-spring or ardvin of a river dedicated to Anahita was at an elevation of 1000 times the height of a man, or 5,700 feet. (Aban Yasht 103) At a somewhat lower elevation, the prophet Zarathustra prayed to Anahita beside the river Daitya, in the Airyana Vaejah, which may indicate he prayed beside the Aras or the Kűr river.

Yasht XXIV:104

> *Unto her (Anahita) did the holy Zarathustra offer up a sacrifice in the Airyana Vaejah, by the good river Daitya.*

Actually, the Kür river rises in Georgia. The Aras is a tributary. What went on at these ceremonials? Here is part of the Aban Yasht, verses 16-17 in case anyone wishes to repeat this performance to get a wish granted!

"Offer up a sacrifice, O Spitama Zarathustra, unto this spring of mine, Ardvi Sura Anahita..."

There follows an account of the wonderful attributes of Anahita...

"To her did the Maker, Ahura Mazda, offer up a sacrifice in the Airyana Vaejah, by the good river Daitya, with the haoma (a drink) *and meat with the baresma* (besom), *with the wisdom of the tongue, with the holy spells, with the deeds, with the libations, and the rightly spoken words"*

Sounds like fun.

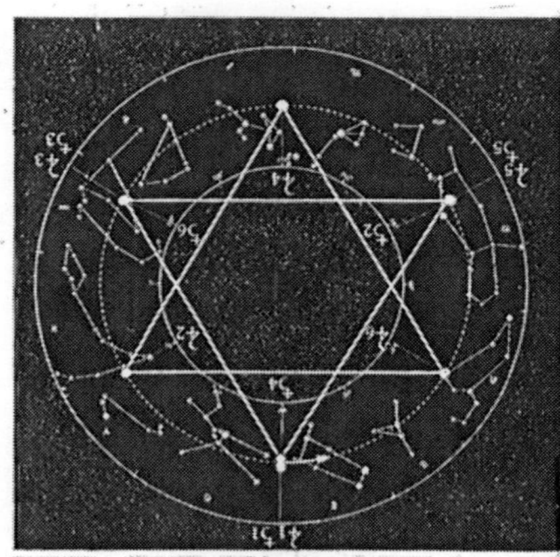

Note: Earth also has Keshvars. They are called the tectonic plates.

ADAM AND THE BOOK OF GOLD

When security lies in secrecy
Safety lies in silence

Anon.

ADAM AND THE BOOK OF GOLD

The first named inhabitant of the Garden of Eden in Paradise was of course Adam together with his wife Eve, over 6,000 years ago, at the beginning of the present 7,000 year cycle according to Hebrew reckoning. There are several texts about Adam, not counting the story told in Genesis, which we will avoid because it is so well known as not to bear repetition. Of greater interest in our context are the texts 'The Prayer of Adam', contained in the work 'The Book of Raziel the King', and 'The Book of the Vestment'. Here we avoid the real Hebrew names of these texts which can be found in our bibliography in the appendix to this book. In the latter work, there is an instruction to "form divisions in the Garden of Eden", but no reason is given. The nature of these divisions will be clear to the reader who has glanced at the illustrations provided.

On then to Adam, who receives a holy 'Book' while in the Garden of Eden which contains all knowledge and wisdom. This was incomprehensible in the past until the advent of computer chips which will soon become so small that a whole library could be contained in the space formerly occupied by just one book. Alternatively, the object given to Adam could have been simply a hand held terminal connected by a radio(?) link to a mainframe elsewhere, possibly on Earth but conceivably in orbit or on some other planet. Taking one more step that we are not yet capable of, the object could have been a thought-reader; very useful where the donor was dealing with a person such as Adam who could not be presumed to be able to read or write... but he could think! Thus, the interaction between this book and a human could be instantaneous, pictorial and of mind boggling scope. At 4000 BC its origin would have to be of alien provenance.

It is quite probable that Adam could make neither head nor tail of this thing, and so in one story proceeded to copy the inscription it bore on to a gold bar he had prepared for that purpose, which could have been done with stone tools, but then hid the gold bar in a cleft in a rock. Brilliant. If there is a vestige of truth in this version of the story, what happened to the original book?

In the work called 'The Prayer of Adam', the book remains a book and is likewise hidden. In another story, the 'book' becomes a sapphire stone, and there are sapphires in Armenia as it so happens. More than this, we are told by records originating in the 18th century AD that the surface of the sapphire was either white or black, just like a computer screen, and the information it contained was displayed in letters of fire. Nothing could be made of this information until the discovery of the liquid crystal, plasma displays, and computer screens in the 20th century. Some researchers think that we only found out about transistors and integrated circuits from downed alien spacecraft recovered by the Americans and reverse engineered, but that remains hypothesis until proved otherwise.

The last time anyone saw this 'Book', was in the 18th century when the story goes that it was in the possession of the 'Baal Shem Tov' (Lord of the Good Name, Yisroel ben Elieze, 1698-1760) living in Eastern Europe in what is now Ukraine, the founder of the Hasidic sect of the Jews, and it may be that he found out how to open its secrets. Ben Elieze is said to have been given the book by Adam the Tzaddik's son when

- this Adam died, by bequest. If anyone is still looking for the secrets of the so-called Illuminati, first find the sapphire stone, or the 'Book'. In case it might be thought that the whole story of Hasidic Jews &c is a fabrication, the Lord of the Good Name must have had a pure soul, because as early as the third or fourth century AD it was written in the "Book of Mysteries" that only the pure could ever gain access to the Book, which would be granted by the angel Hadarniel; the angel Raziel presumably having died which after the passage of 2000 years seems a certainty. It seems to be an easy deduction that the object, whatever it is, passed into the hands of Adam Weishaupt (1748-1830), the founder of the Bavarian Illuminati, and from him to certain Americans around the time of the declaration of independence (George Washington? 1732-1799), and thence the New World Order as it is now known. Hitler may have known of it but never found it despite desperate attempts to gain occult power by collecting sacred relics. It must be noted, however, that any Grail search is a dead end to illumination.

It is of interest that sapphires can display 'asterism', or the ability to show six rays in the form of a star, when cut and polished in a particular manner, so as to 'reflect' the divisions of the Garden of Eden, itself arranged as a six pointed star. An aster is a daisy, but also means 'a star', and if this is constructed within a circle using a pair of compasses, a diagram results that can be found in Egypt and Armenia, not to mention countless rose windows of churches and cathedrals made by masons who knew what they were doing. In fact, the divisions of the Garden of Eden are displayed everywhere and the masonic last laugh is on the peasant who gazes in wonder and understands nothing. You may, if you choose, have twelve divisions of the circle to count the hours, but six are redundant.

Quite often, the twelve divisions of a rose window contain illustrations of the twelve apostles or disciples of Jesus, or perhaps twelve saints..

The personage who gave the Book to Adam was, in the Prayer of Adam, the angel Raziel who is sometimes incorrectly known as Rezial, but then vowels were always considered of little importance in the original abominable Hebrew alphabet, which ought to be compared with the Armenian alphabet if justice were to be done.

Raziel was no angel; he was a walking talking human even if we accept that he was sent by God. Part of this story is told by Erich von Dåniken in his "Return of the Gods" (1995, 1998 Element, UK) page 25 et seq and more is given in the "Sepher Rezial Hemelach", details of which are in the appendix.

"*I* (Adam) *pray you* (God) *will reveal what will come of the generations coming after; 'what will occur every day and every month. I pray you do not conceal the wisdom. Watch over me and sustain my labours*".

Adam prayed for three days. God sent forth Raziel who dwelled upon the river (Aras) *going forth from the Garden of Eden.*

He (Raziel) *was revealed to Adam as the Sun went black. By his hand, he gave the book to Adam saying: "Do not fear and lament any longer. From the day you served in prayer, the prayers were heard. I come to give you the knowlege of the words of purity and great wisdom. Become wise by the words of this most holy book...every word is revealed".*

Adam drew near and heard, learning to be guided by the holy book. Raziel, the angel, opened the book and read the words. Hearing the words of the holy book from the mouth of Raziel the angel, he fell on the ground trembling in fear.

Raziel spoke saying: "Rise up and be strong. Revere the power of God. Take the book from my hand and learn from it. Understand the knowledge. Make it known to all who are pure. Thereby establish what will occur in all time."

Adam took the book. A great fire kindled upon the bank of the river. The angel rose up in flames and returned to Heaven.

If one prefers the alternative version, it is written:

"Adam prepared a bar of gold and purified it twice by fire. Then (he) engraved therupon the signs of the holy book, in holiness and purity. Then (he) placed the book in a cleft of the rock on the east side of the Garden of Eden- not to fall to earth but to endure and sustain the power. In every place, Adam was sustained. He wandered upon the Earth as a ship upon the sea. He remembered and recorded when mountains melted like wax; and the sea went forth as hoar frost, and fire went forth as dust, and lions and leopards and bears and animals of the pasture stood silent and still; and insects rose up in the air as birds; and the sea became a desert wilderness....He recorded when the stone wall of the tower suddenly fell to the ground; and fruit trees bore fruit before the proper time; and peoples take flight before the battles of war."

This passage is not put in italics because some or all may be a gloss on the original and is taken from 'The Book of the Vestment' (whatever that means).

This tower is mentioned in the Book of Isaiah in connection with a messiah, who might have been supposed to arrive after the twin towers fell on 9/11 but did not. A depiction of the tower exists in the tarot pack where it is struck by lightning.

Now Adam concealed the book/bar in a cleft in a rock but it served his son Seth for ten years and he too kept it hidden away in a cleft in a rock, or in another story, in a golden vault kept in a cave. It remained there for four generations until it was found by Enoch, the son of Jared, as the consequence of a revelation shown him in a dream. After Enoch was 'taken by God' the book remained hidden until the arrival of Noah, the son of Lamech, who was presented with this fabulous item by the angel Raphel at the command of the Elohim.

Noah was guided by the wisdom of the book and made it known to his son Shem, who handed it on to Abraham; passing in time to Isaac, Jacob, Levi, and others down to Moses (Moyshe). The book was buried with Joseph in Egypt but was dug up and retrieved along with Joseph's remains to be carried through the wilderness of Sinai alongside the fabled Ark when the Hebrews escaped from Egypt with the intention of reaching the Promised Land, which was not Armenia.

The book passed into the possession of King Solomon who may have used it to read instructions on how to build a temple, but Solomon's temple has been almost completely destroyed save for a famous wall. The book was then deemed lost, but it was said that God had instructed the angel Hadarniel to conceal it once more in a cave, allowing only men with pure souls to see it. That was known for over a thousand years...

ARMENIA TO NEW ZEALAND

According to a survey, one third of all UK adults think that Mount Everest is in Europe.

Sky TV News, November 2007

According to rumour, one third of all Australians think that New Zealand is an offshore island just over the horizon near Sydney.

'Om' magazine, 1981

ARMENIA TO NEW ZEALAND

It may be asked why on earth should anyone want to voyage halfway around the globe, as for instance known sites in Australia that were visited by ancient Egyptians, or the Great Lakes region of Canada known to the Phoenicians, but the answer is simple; metals, spices, furs, slaves and general trade and plunder. Most of the voyages in antiquity took at least one year and more likely two, because the Trade Winds had to be used both there and back, while crops might be grown while waiting for the wind to change.

There were many occasions when all that mattered was escape, as for instance when an enemy fleet was bearing down on you or a nearby volcano was about to blow its top. If one did not escape, the consequences would include being buried alive, famine, disease, chaos, and death, and it is likely that the escapers harboured no thoughts of returning, becoming 'Uruk' or colonists.

Some significant dates when humanity was overwhelmed by Nature's paroxysms are 3237 (Noah's Flood), 2345, 1628 (Santorini), and 1453 BC (The Plagues of Egypt) but if you lived in New Zealand you might not have survived the eruptions of 1320 BC (Waimihia), 1100 (Kaharoa) and the 186 AD eruption that blasted out the crater that holds Lake Taupo.

In general, if it is not famine or warfare that prompts your permanent migration, it has to be fear of the repetition of some celestial event that wiped out your ancestors, in mitigation of which you will make sacrifices and offer prayers. The mere sight of a large comet might be the impetus to migration- in many instances we simply do not know, but here we shall examine what happened to the peoples of Urartu, once the name for the whole of Armenia.

In the first millennium BC, the land of Armenia was called Urartu, centered on Lake Van, and covering all of present day Armenia and some of eastern Turkey as well. Urartu was supposed to have been founded by Aram, one of the first Armenian patriarchs and existed from 810 to 714 BC. The whole land was then called the Airyana Vaejah, which is taken from the Zoroastrian, meaning 'The first and best of all dwelling places'.

The colonists who left the Airyana Vaejah, or Paradise, called themselves the Aria or Arya, Airi, Ide, or the Ariyan people- Aryans in English of course.
Remembering that Uruk means colonist, it is quite remarkable that a tradition exists in New Zealand among the indigenous Maori that their ancestors, the *Urukehu,* came from *Airihia*. Others called the 'Kiripuwhero' came from *Uru* (Urartia), according to students of the 'whare waananga', the House of Initiation or Tradition where everything taught has to be memorised. A New Zealand historian, familiar with such matters, states that her ancestors prohibited writing no less than four times, and so the Maori had no writing until the 'Pakeha' (White Man) arrived in the early 19th century. This is a familiar prohibition, as for example among the Druidic Celts who shared the same thinking regarding memorisation of the secrets only shared with the worthy in the first place. Celt is 'ceilt', or hidden, in the Celtic language, whose traditions originated from the same area, Turkey, the Black Sea, Georgia, the Ukraine, and Urartu.

Author Elsdon Best says that Airihia is an ancient raised homeland, now called the Caucases, and this opinion is shared by Elders (1975 and 1976), by Edward Tregear, and 'Mana' who writes for www.celticnz.co.nz. where this info is to be found.

Tregear says that the *'reo'* or language of the Maori retains the purity of their part-Aryan forebears, showing that there are 2800 words in Greek and Maori with the same pronounciation and meaning! This is in spite of an ancestry that includes possibilities such as Taiwan and Melanesia, as well as Poynesian origins. In fact, Hawaiian is comparable with Maori although the two are geographically thousands of miles apart, as is Fijian, roughly halfway.

The Maori know that the early voyagers were a forestry people, and had achieved a high degree of spiritual understanding. They say the Uruk used the Bosphorus as their exit from the Black Sea and all that is left of their history is the most recent fifth, still held by the whare waananga.

In summation, the timescale presented here means that the Maori spent roughly 2000 years completing their exodus from Urartia to Aotearoha, as they call New Zealand.

When the exodus to the East from Armenia began, those who would become Maori, which means 'us', were *white Aryans* who travelled into India and China. From India, they would have sailed towards Indonesia and then Melanesia, finding the northern tip of Australia before continuing into the Pacific Ocean where some became Polynesians. Meanwhile, those who had chosen China reached Taiwan before sailing south where their genes mingled with Phillipinos, and so Maori ancestry includes Taiwanese, but it seems that the ocean navigators remained a distinct clan who would have preferred to intermarry with their own kind, no matter where they had voyaged from. This preserved the Polynesian language which is basically understandable from Hawaii to New Zealand and everywhere in between.

The last port of call of our Maori was Raratonga, to the north-east of New Zealand, which supplied at least seven and maybe 10 waka (canoes) carrying several hundred people to their new home, which they must have known about before setting off. This was no desperate semi-suicide trip, arguing that New Zealand *was known and visited long before.*

HE IWI TAHI TATOU

We are now one people.

Te Arero, NZ.

HE IWI TAHI TATOU

In this chapter, we present some interesting email correspondence between two contributors to the website **www.celticnz.co.nz** where they discuss the origin of certain New Zealand Maori tribes, called *iwi*. It is our opinion that *iwi* is the same as *iki*, and that the legendary Maori homeland of *Hawaiki* has the same meaning as *Hawai'i*, where the 'k' has been dropped as is very common in some Polynesian languages, resulting in almost incomprehensible syllables such as *a'a* and *o'o*. Hawaiki is in fact a name for some place that the Polynesians vaguely remember as 'home', but they have forgotten exactly where it was. The authors of www.celtic.co.nz are more specific in stating that their remote ancestors came from *Airihia*, which we know of as the Airyana Vaejah.

The degeneration of language in the Pacific was matched by the loss of most of their original culture, and soon they reverted to a stone age society. Here we may mention that it is our opinion that the Maori did not carve their cherished greenstone ornaments but stole them from previous residents who they had murdered, being unable to carve their own jade objects out of this remarkably tough material.

The New Zealand authors of www.celtic.co.nz concur that the Maori, who arrived some time between 900 and 1300 AD, found other peoples in residence who should properly be called *tangata whenua*, the Maori words for the 'first people'. However, after the *pakeha*, or white man arrived, the Maori started calling themselves the *tangata whenua*!

These correspondents go further than usual in stating, perhaps for the first time in public, that they claim ancestry of the *Uru*, who were in all likelihood the Urartians of Armenia, and the *Urukehu*, who were no less than the 'Uruk' or colonists from the same area of the world. Their migrations and explorations started around 900 BC and so the transformation of Urartians to Maori (meaning 'us') took no less than 1800 years. Note that the word 'Maoris' is inappropriate since there is no plural of 'us', nor is the Maori equivalent *'nga Maori'* ever used.

Here, then, is how the two correspondents describe their history-

"I am Iwi Maori and I am also Tangata Whenua. When my ancestor Ngatoroirangi arrived upon (the) Te Arawa waka (canoe) he found earlier people here including Hapuneone, Maruiwi, and even Patupairehe ('fairy folk'). These peoples were not unknown to him because they too came from Hawaiki in earlier migrations- note that it was Ngatoroirangi who brought the Patupairehe people to the inland region of Te Matapuna! These peoples had much intermarriage with each other and ALL of them are my ancestors.
Yes my ancestors are Urekehu from Airihia and Kiripuwhero from Uru. My ancestors are Kiritea from Arinuiteao and also Hema; ALL these people came together to create the Tangata Maori- the peoples of Polynesia. My ancestors came from the land Airihia which is also called Arya.

According to the (Zend) Avesta holy text of Zoroaster, Aryana Vejahe is the home of Aryans, the Airihia, and later it was called Bactra and then Khorasan. The Aryan civilisation of my ancestors encompassed a large area, at the heart of which was the land of Khorasan that covers today's Afghanistan, eastern Iran, and central Asian countries.

These are the people whose mummified bodies are found in China, and who crossed into America also. Polynesian tangata whenua and American tangata whenua all derive from Asia, and before that back to Airihia, according to true human lines from Eve in the Garden of Africa-Eden (sic). From here too they moved towards Europe and there encountered the Neanderthal inhabitants of that peninsula.

I am tangata Maori, a human being, of the Tuwhatetoa iwi and also tangata whenua through intermarriage between Hapuneone and Nga Oho of Te Arawa waka."

Te Arero, NZ.

Apologies if the reader is not given explanations of every word but space is limited as is the Editor's knowledge! We continue with the reply to the above passage-

"Indeed, some of the early groups had formerly lived throughout Oceania, but because of warfare had to seek sanctuary in New Zealand or elsewhere. One of these early groups was the 'Children of Poutini' who were the offspring of Tangaroa, the god of ocean migration. Tangaroa was described as 'sandy haired' and had originally come from a land where leaves stayed on the trees for half the year. His offspring were finally chased from the isles of Oceania, including Tonga and Samoa, never to return.

From Easter Island to Tahiti and Tonga, as well as further afield, these 'stone builders' left many temples and other large cut stone/stacked structures or paved roadways/pathways, artifacts, &c. The island of Moorea in the Tahitian group is known in regional oral traditions as 'the island of the fairy folk with golden hair' (Patupairehe, Ed.) not unlike many descriptions of the small stature Turehu of New Zealand.

The 'Children of Poutini' , amongst many others found by your Maori people upon arrival from Hawaiki (probably Raratonga, Ed.) were the great carvers of 'pounamu' (Greenstone, jade) and are thus referred to in Maori tradition as 'the greenstone folk'. Of the Children of Poutini, Maori oral (is there any other? Ed.) tradition states: They were attacked here and lost a number of their people who were slain, which was a benefit to the Maori people. These slain and captured Greenstone Folk (are) represented (by) blocks of greenstone aquired by the Maori (including) famous greenstone artifacts.

The Children of Poutini, like the 'Moon Maidens' who were the offspring of Tangaroa, were 'urekuhu' (having) reddish, golden tinged hair, and 'kiri puwhero' reddish, light complexioned skin. This was an era when war, strife, quarreling, were brought here by the people who came from Hawaiki."

Mana, NZ.

The reader may remember that the gods of Armenia were likewise light complexioned, with fair curly hair and blue eyes, and it is doubtful if any Maori would know that. Today, we might term such people 'copper knobs' who burn in the sun and might come from Scotland. Again, a long way from Armenia, and of course, anyone from Mars would not be used to the brighter Terran sunshine, now would they?

HAWK NOSED HEROES

Eternal growth is cancer.

Piachi Diddicoy

HAWK NOSED HEROES

The alphabets used today, no matter which one we consider, seem to have a common origin, sought by learned men throughout antiquity to the present day. Their search for the 'mother tongue' led them to Sumerian and Babylonian scripts, to Egyptian heiroglyphs, to Crete, and many other systems of writing back to petroglyphs from tens of thousands of years ago.

An Armenian would say that the origin of the alphabet was *yergatagir*, the Iron Age script used at 2000 BC and traceable back to the indigenous 'Armenid' Subarians of 4000 BC. They were conquered by the incoming 'Aryan' Haiasians, whose semi-legendary king *Hayk* gave his name to the land of *Haiastan*, whose inhabitants are called *Haikarian* to this day. The 'k' of Hayk may be dropped, producing a name rendered as 'high' and basically meaning husbandman.

The northern tribes became the *Haiasa-Azzi*, or the *Hyksos*, also known as the Shepherd Kings, who created one of the world's first empires including Syria, Egypt, and Palestine, who gave the yergatagir script to all these places which would eventually out-last the Hyksos, who ruled from 1730 to 1580 BC.

They were responsible for the introduction of the horse and chariot, bronze and iron weapons, scale armour, the war helmet, the composite bow, a complete calendar, and moated cities. New and better methods of spinning and weaving were introduced to the 'Semitic' world together with new musical instruments, the hump-backed bull, the olive and pomegranate, and set the seeds of monotheism by worship of Ar, the Sun. Skeletal remains and anthropological evidence of the past and present attest that in the Armenian Highlands, there evolved an indigenous racial type with unique physical and mental capacity. This race, known as the Armenoid or Armenid is described as characteristically sturdy, often long-faced, abundantly bearded, with a backward sloping forehead, receding chin, and adorned with a prominent aquiline nose. Interestingly, many of traits are noted among the remains of Cro-Magnon Man, the ancestor of Modern Man, while the ancestors of the White Race are termed Caucasian, or the inhabitants of Georgia and Armenia. Wherever the Armenids went, the locals tried to imitate the features of these Caucasians by artificial deformation of the skull, practiced on new-born babies by binding with cords and wood. This spread all over the world; to Asia, Africa, Europe, and America, among other races widely separated in space and time, as for instance the Celts, Etruscans, Anglo-Saxons, Swiss, Scandinavians, French, Scythians and Turks, characterised by the prominent 'Semitic' nose.

The ruling classes of the Turks and Egyptians were likewise Armenid, but further away in Melanesia, Polynesia, the Pacific Islands and especially the negrids of New Guinea, the prominent hawk nose is a testimony that the Armenids were present and numerous enough to have left their genes in these places.

Surprisingly, it was amongst the peoples of the Americas that the hawk nose and head deformation was most venerated; as seen among the Aztecs, Inca, Maya, and some of the North American tribes such as the Chinook, Clatshop, and Salishan. Among the Algonquin and Blackfoot tribes, the intermarriage was such that the hawk nose came about in these tribes by natural birth. This was a mark of aristocracy that endured from the Cro-Magnon to Roman days in Europe.

The impetus to this world-wide conquest was initially the search for metals; especially copper and tin, causing prospectors initially from the Caucases to push their way into the Balkans, the Danube Valley, central Europe, Denmark and Sweden. It was a short hop to coast of England where tin, lead, and copper were available, and from there we may assume that they crossed via the island of Angelsea into Ireland where the peat bogs had not yet completely covered the extensive deposits of gold. Their most remarkable achievement was to cross to Canada in North America where millions of pounds of native copper were extracted from the Great Lakes region, and one does not move that quantity of metal by St. Brendan type skin boats.

Others, somewhat earlier, had followed the obsidian trade routes southwards into Syria, Mesopotamia, Palestine, and Egypt, culminating in the Phoenicians or 'Sea People' who also displayed the hawkish nose.

Nearly all these folks mentioned had writing, although due to warfare, the original yergatagir script of Armenia was almost lost, although there is one direct link between this old alphabet and the modern (5th C.) alphabet invented by St. Mesrob Mashtots.

In 1830 AD, in the Armenian monastery of St. Lazar in Odessa, Russia, there existed an ancient Armenian leather-bound book which had originally been found in the Sinai Peninsula. Today, only two pages survive out of 32, dated between the 5th and 4th centuries BC; this book is the only link between the script of the Hyksos and the Mesrobian letters. True to form, similar scrolls have been found by Dutch excavators working in Jordan but they have not been published. Could it be that history would have to be rewritten?

Those who have followed the narrative in this book with a measure of disbelief may entertain the thought that all, repeat *all*, of the stone circles of this world; its monuments which are still standing in the image of the Vishap stones; its sighting stones for the calculation of the equinoxes, and the rest of the impossible stonemasonry from the Poles to the Pacific is the result of a culture spread by Armenians who should be given the credit they deserve.

They advanced humanity from the Stone Age to the Age of Technology and it could not have been accomplished without writing of one kind or another.
Please do not continue to call the original language of Earth 'Adamic' or 'Nostratic' because now you know the real name of the oldest script and language on this planet. It is of no matter if the reader is prepared to believe that the initial civilising influence was 'off-world', even Martian, or whether some destroyed race such as the hypothetical Atlanteans resurrected their world in Armenia, or if some other stranger hypothesis is entertained, the fact remains that answers can be found here, if we wish.

It was here that humanity progressed from darkness to mental and spiritual enlightenment. It may sound fantastic that such a small population as existed in the fastness of Caucasian valleys could so influence the world but they did. It is not their fault that we have 'progressed' to nuclear weapons, universal instant communications, supersonic flying machines, an oil based culture that demands total mobility for everybody despite the obvious crazy and catastrophic consequences. We have gone too far. Soon, another planetary-wide catastrophe not of our making will send us back to square one with the difference this time being that the resources of Earth have been stripped.

Should Man-made global warming overtake our ability to prepare and deal with it, then we may cook in our own juices before the coming natural catastrophe on its way from the furthest reaches of the Solar System turns the thermostat to 'very cold and erratic'. It may be that if the worst possible scenario comes to pass, then the Earth may be turned upside down, volcanos will run riot, and gigantic earthquakes will rearrange the geography of the tortured planet. In time, things would settle down and a new world of fresh possibilities could be said to have emerged from ghastly chaos, just as the oldest legends and newest prophecies of our planet tell, re-tell, and fore-tell.

Then, somewhere on Earth, a tiny band of humans will start again, somewhere safe, hemmed in by mountains, with the prospect of growing crops, a permanent supply of clean water, and a supply of metals and some firewood or coal to hand... does this sound familiar?

Note: several quotes from the work of Melkon Armen Khandjian are incorporated in this chapter, for which the Editor tenders his sincere thanks and appreciation. Permission to quote was sought but no answer was received. We print nevertheless.

APPENDIX

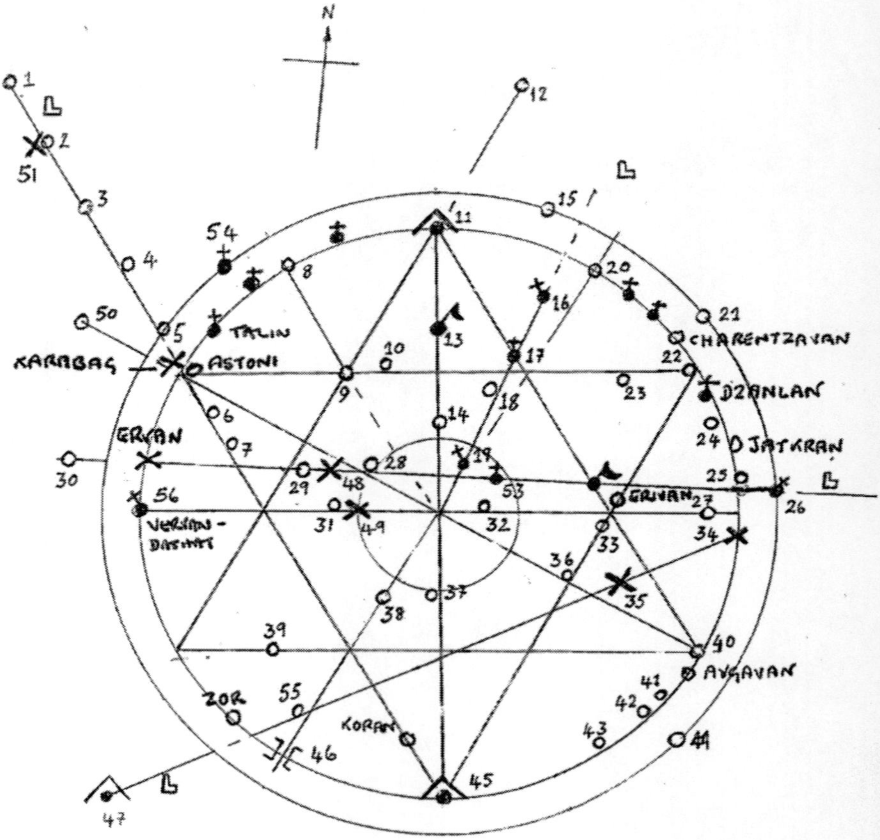

THE YEREVAN STAR

KEY:

- mountain
- fort
- town or village
- church
- pass
- antiquity (site of)
- ley-line

MAP SHOWING ADDITIONAL SITES

1 Esenkent
2 Ocakir
3 Anipenza
4 Akko
5 Arteni
6 Karaker
7 Bagramjan
8 Kamrasen
9 Arue
10 Kos
11 Mt Aragats
12 Alagjaz
13 Hamberd Fort
14 Achnavartun
15 Kucak
16 Sarnosavank Ch.
17 Karbi & Ch.
18 Astarak
19 Ecmiadzin Ch.
20 Buzakan
21 Carencavan
22 Kaputan
23 Nor Acin
24 Zar
25 Geghard
26 Geghard Ch.
27 Garni
28 Mecamor Antiquity
29 Hoktemberian
30 Bagaran
31 Armavir
32 Dostlug
33 Sovetasin
34 Garni Antiquity
35 Dvin Antiquity
36 Mchcran
37 Tasburun
38 Karakoyunin
39 Halfeli
40 Bedi
41 Ajgavan
42 Avsan
43 Igdire
44 Ararat
45 Mt Ararat
46 Cilli Gecidi Pass.
47 Mt Akpinir Tepesi
48 Mecamor Antiquity
49 Armavir Antiquity
50 Aragac
51 Ani Antiquity
52 Erebuni Fort
53 Svatnoz Ch.
54 Vreriuk Ch.
55 Suveren
56 Vervan-Dashat

All places are villages or towns unless otherwise indicated.

An Antiquity is is an ancient megalithic site

Spelling may vary from map to map. Alternatives are not given here.

from an original by Vancouver Travel Maps.

The additional circles shown on this version of the Star in Paradise are occasionally shown on other diagrams and can frequently be seen on Jewish grave headstones. Comparisons may be made with the earlier diagram (The Gates of Eden) shown in this book, and extended ley-lines in the first chapter.

OF DILMUN AND GILAN

Dilum, Dulum. Le Strange, writing of the alluvial delta lands of the Safīd Rūd on the Caspian, called Jīl or Jīlah by Arab geographers (see *Gilan*) writes :—

" To the south and west " (viz. of Jîlanât) " the mountain range bordering on the districts of Ṭâliḳan and Ṭârum in the Jibâl province was the Daylam country, generally also given the plural form as Ad-Daylamân ; and this country became famous in history as the original home of the Buyids or Daylamites, whose chiefs were masters of Baghdad, and of the Caliphate for the most part, during the 4th (10th) century. . . . . When Muḳaddasi wrote in the 4th (10th) century, and the Buyid supremacy was at its height, all Gilân, together with the mountain provinces to the eastward and along the shore of the Caspian, namely Tabaristân, Jurjâh, and Ḳumîs, were included in the province of Daylam, but in later times these eastern provinces came to be counted as separate. Afterwards the name of Daylam itself for the most part fell out of use and the lowlands of the Sâfid Rûd delta gave their name to the whole of the adjacent district, which was commonly known as the Jîlân province. More exactly, however, Jîlân was the coast district, while Daylam was the mountain region overhanging it, and at different times either of these names in turn might be taken commonly to include the whole province lying round the south-western corner of the Caspian Sea. . . . To the westward of Rasht extends at the present day the district of Tûlim, and Mustawfî gives this as the name of an important town in the 8th (13th) century." (Le Str., chap. xii.)

Professor Minorsky agrees that the Dilum of our MSS. may be Dailam. He writes : " *Delymais* is found already in Ptolemy." He gives as a second possibility that the name may refer to the Babrain Is. called Dilmun in ancient times.

Mr. Theodore Gaster also suggests Dilmun in the Persian Gulf, " often mentioned in Babylonian texts," and gives reference to Fr. Belitsch, *Wo lag das Paradies*, p. 229, where the name DALAMU occurs.

There remains a possibility that it was the ancient Dolomene. Strabo (Str., vol. vii, p. 193) writes :—

" The country of the Assyrians borders on Persia and Susiana. This name is given to Babylonia and to much of the country all round, which latter, in part, is also called Aturia, in which are Chalontis in the neighbourhood of Ninus, Appolloniatis, the Elymaei (Elamites), the Paraetacae, the Chalontis in the neighbourhood of Ninus, and also Dolomene and Calachene and Chazene and Adiabene, and the tribes of Mesopotamia in the neighbourhood of the Gordyaeans, and the Mygdonians in the neighbourhood of Nisibis, as far as the Zeugma of the Euphrates, which is occupied by Arabians, and those people who in a special sense of the term are called by the men of to-day Syrians, who extend as far as the Cilicians and the Phoenicians and the Judaeans and the sea that is opposite the Aegyptian Sea and the Gulf of Issus."

Dilmaiia. The Daylamites, the people of Dilum. See *Dilum.*

ARMENIAN FIRSTS

Years BC	
500K - 400K	Abbevillean or Chellean Culture. Earliest stone tools in the Former Soviet Union (Armenia/Georgia) on the SW slopes of Mt Aragats (AlagozAragac). Major complex of workshops near Kisir Dagh or Arpa Chai Range.
400K - 100K	Acheulian Culture. Obsidian and dacite knives scrapers, hand axes, spear heads, darts, spikes well represented suggesting continuous habitation. Main centers were the Ukraine, Crimea, and Turkmenia on the shores of the Caspian Sea.
100K - 40K	Middle Paleolithic, or Ice Age. Neanderthal still in existence making side-scrapers, disc-cores, flake tools, knives, spear heads, arrow heads. Important centers of Neanderthal/ Mousterian Man were Georgia, Armenia, and Azerbaijan. Modern Man gradually displaces Neanderthal, diffusing and improving on previous cultures.
40K - 12K	Emergence of Cro-Magnon, Ice Age still evident and named Upper Paleolithic, Lower Aurignacian or Chatelperonnian. Latter develops new kind of stone knife including keel and fluting, probably invented in the Caucases and export -ed into Western Europe, the Sea of Azov, the Don and Volga.
10K - 6K	Magdalenian in evidence, semi-nomads of Europe.

HISTORY AFTER THE LAST ICE AGE

BC	ARMENIA		NEW ZEALAND
8000	Cold working of metal		
7000	Pottery		
6500	Obsidian trading		
5000	Smelting begins		
4000	Armenid Subarians Medzamor constructed		
3000	First observatory Invention of chariot. Horse tamed		
2500	meteroric iron used		
2000	Aryan Haiasians Hattians Hurrians Mitanni 'Yergatagir' or Iron Age script		
1730	Hyksos Empire starts		
1580	Ends	c.1300	First known humans
1400	Steel production begins	1320	Waimihia eruption
1200	Nairian Empire	1100	Kaharoa eruption
900	Urartian Empire	AD 186	Taupo eruption
		c. 1250	Maori arrival

THE BIBLIOGRAPHY

'Return to Mars' Crowley and Pollock, Magistra, Vic. Australia 1989 and Black Rabbit Press 2003.
'The Sand Whales of Mars' McLeod and Austin, Black Rabbit Press, UK, 2001.
'Dictionary of Symbolism' C.E. Cirlot, London, 1962.
'The Dragon's Tail' Austin and Crowley, Illuminet, Ga. USA 2000.
'Draco: The Tenth Planet' Austin & Crowley, Black Rabbit Press, UK 2003.
'Hanes Taliesin' (The Book of Taliesin) J.E. Caerwin, Lond. 1968.
'The Key' Cohane, UK.
'The Proto-language' Patrick C Ryan, 1998, UK
'The Book of the Zodiac' trans. E.S. Drower, Royal Asiatic Soc. London, 1949. re p. 207.
'Sacred Books of the East' trans. J. Darmesteter, American Edn 1898 ref. 'Fargard 1' and Yashts.
'Sacred History Research Expedition Notes' Col. Alexander A. Koor (pseudonym) 1946 re carvings at Karada (Kara Dagh).
'Wir, vom Mars' (We, from Mars) Walter Hain, Vienna, Austria 1992, also Ellenberg Verlag, Köln, Germany 1979.
'Gods of Eden' 1998 and 'Ashes of Angels' 1996 A. Collins, Headline, London.
'Ancient Records of Egypt', J.H. Brested, Chicago, USA 1906.
'The Genius of the Few', O'Brien, Turnstone, Wellingboro, Kent.
'Armenia, the Cradle of Civilisation', D.M. Lang, Allen & Unwin, London 1970.
'Mandaeans of Iran & Iraq', Drower, Clarendon Press, Oxford, UK
'The Foundation of the Iranian Religions' Prof. L.H. Gray, KR Kama No. 5, 193?
'Other Tongues, Other Flesh' G.H. Williamson.
?? G. LeStrange, Cambridge Uni Press, UK, 1930 re Dilmun.
'Secrets of the Lost Races' R. Noorbergen, Bobbs-Merrill, 1977.
'Encyclopaedia Judaica' re p.473, Armenia as Paradise.
'View over Atlantis' and 'New view over Atlantis', 'City of Revelation', Mitchell, Abacus, Sphere, and Garnstone Presses, 1972 onwards.
'Ancient Records of Egypt,' Brested.
'Sepher Rezial Hemelach', trans Steve Savedow, Samuel Weiser, USA 2000.
'Rediscover Armenia', Brady Kiesling, Amazon, 1999.
'King James Bible' re Armenia: Gen 10:22, 25:30, 28:5, 2:8, Isa Gen 14:13, 8:4 Eze. 28:13-16 II Kings 19:37, Jer. 51:27, Amos 4:3, the Aggadot of Esther.
'The Tribes', Yair Davidy, Russel-Davis, Israel, 1999.
'Earth Memory' Paul Devereux.
'The New Life Prophecies of Frances Yule', Ed. Venestus, Destiny Pubs. NZ. South Australia, 1982 (v. rare).

'Saucers of the Illuminati' and other works, Jim Keith, Illuminet, Ga. USA 1995 onwards.
'Return of the Gods' 1995 & 1998 and other works by Erich von Däniken, 1960s onwards.
'The Cycles of Heaven' G.L. Playfair & Scott Hill, Souvenir, London and Canada, 1978.
'Georgia, Armenia, Azerbaijan', Plunkett & Masters, Lonely Planet, London & Oakland, Ca. USA. 2000 & 2004.
'Raratonga & the Cook Islands' Hunt & Keller, Lonely Planet, London & Oakland, USA, 1986 - 2003 (5th Edn).
'New Zealand' var contrib. Dorling Kindersley, London, UK, 2001 - 2003.
'Die Chronik von Akakor' (The Chronicle of Fortress Two) K. Brugger, Dusseldorf & Wien, 1976. Also an English translation.
'A dictionary of Angels' G. Davidson, London, 1968.
'The Lost Ship of Noah' C. Berlitz, New York, 1987.
'Someone Else is on our Moon', G. Leonard, London 1977.
'Intelligent Life in the Universe' C. Sagan & I.C. Shklovski, San Francisco, USA, 1966.
'Destiny Mars', M.W. Saunders & D. Lunan, Caterham, UK 1975.
'The Other Atlantis' R.J. Scrutton, London, 1977.
'We discovered Alien Bases on the Moon', F. Steckling, Vista, Ca, USA, 1981.
'The Sirius Mystery' R. K.G. Temple, London, 1976 (repub.)
'The Pictorial Key to the Tarot' A.E. Waite, London, 1910.
'Our mysterious Spaceship Moon' 1976, and 'Secrets of our Spaceship Moon' London 1980.
www. mars-mania (Keith Phillips, USA) re Mars &c.
www. enigmasmania (Keith Phillips, USA) re Mars &c.
www. celticnz (I.O.Man? NZ?) 100+ page correspondence on Maori.
www. tourarmenia (Arm.) re history of Armenia.
www. saintsarkis.org (USA) re: 'The Evolution of Armenian and World Alphabets', M.A. Khandjian, 2001 - 2007.
www. prophecies.us re: articles by tonyaustin 2006 on.
www. mars-news.de (Holger Isenberg, Germany).
Vancouver Travel Maps, Canada.
Malin Space Sciences Systems (MSSS), USA, re: Mars photos.
'Nexus' magazine, Qld. Australia, also UK. 1990s on.
www.theserenedragon.net re dragonology.
www.projectdiscovery.net re Vishap stones.
'Book of Mysteries' (Dead Sea Scrolls) O. Piper, Journal of Religion 38, 1958 pp. 95-106.
'Necronomicon' trans Olaus Wormius 1487, latin edn. from the Arabic. 2) 'The Wanderings of Alhazred' Donald Tyson. (Amazon).

Printed in the United Kingdom
by Lightning Source UK Ltd.
132508UK00001B/178-246/P

9 780954 029845